D0894310

Elements of Entomology

Elements of Entomology

An Introduction to the Study of Insects

Harold Oldroyd

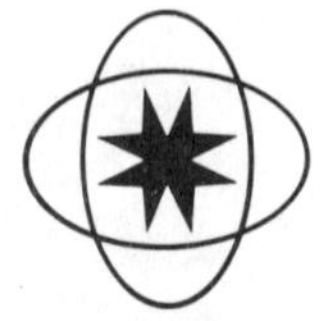

UNIVERSE BOOKS
New York

Published in the United States of America in 1970
by Universe Books
381 Park Avenue South, New York, N.Y. 10016

Library of Congress Catalog Card Number: LC 77-116574

ISBN 0-87663-127-8

Printed in Great Britain

Contents

List of plates

Acknowledgements

The publishers thank the following for permission to reproduce photographs in this book: Walther Rohdich, plates 1, 15, 32; A. E. McR. Pearce, plate 5; Fritz Siedel, plates 8, 25; L. W. Brownell, plate 9; Raymond A. Brabbins, plate 10; Richard L. Cassell, plates 13, 18, 21, 29, 30, 31; Pitkin, Bildpost, plate 17; J. J. Ward, plate 19; G. Ronald Austing, plate 20; Harald Doering, plates 14, 35, 46, 47; Treat Davidson, plates 22, 23; Robert C. Hermes, plate 33, Leonard Lee Rue III, plate 34; Lynwood M. Chase, plates 36, 38; Fred H. Wylie, plate 37; Edwin Way Teale, plates 39, 45; Hugh Spencer, plate 43; J. M. Hood, plate 44; M. Mihara, plate 48; H. Bäufiger, plates 49, 50.

Preface to the American Edition

The appearance of an American edition of my book gives me a welcome opportunity to comment on some of the observations and criticisms that have reached me since it first appeared in England. Reviewers in general have been kind, and have accepted the book in the way in which it was intended, as background reading, and not as a primary textbook. They have pointed out some typographical errors, some matters on which opinions differ, and a few examples of plain ignorance on the part of the author. I regret all of these, especially the last.

Some reviewers have criticised the amount of space used to describe all the different orders of insects, saying that this should have been left to the textbooks, and the space in this book given over to entirely general topics. I should have preferred to do just that, but I think that an elementary book should start at the beginning, and that it is not profitable to draw comparisons between different groups of insects until they have been defined.

In order to make the book readable I deliberately avoided cross-references from text to figures, and hoped that the captions would explain themselves. The letters and numbers in figures 16 17, 18 and 40 are the usual code-symbols for the wing-veins in the Comstock-Needham notation, and it is not necessary to put them into words: thus the vein labelled R_5 is spoken of as 'R_5'. Figure 1 shows only the hard parts of a segment, and the dotted parts labelled 'phragma' are internal plates or ridges, serving as strengtheners, and as additional attachments for muscles.

Limulus, the king-crab (p. 8) is nowadays removed from Arachnida into a separate Class, but is an arachnid in older textbooks. On p. 10, line 18, Myriapods should read 'Diplopods' (millipedes), and the last paragraph of p. 22 unintentionally implies that Neuroptera are Hemimetabola: in place of 'and Alder-flies', read 'like the larvae of Alder-flies'.

I was apparently wrong to equate chrysalis with cocoon, though it is equally wrong, in my opinion, to equate it with pupa. The Greek word means a golden object, and to say, as one author does: 'the chrysalis is generally green, sometimes grey or brown' is a contradiction in terms. Pupa and cocoon are the scientific terms, and chrysalis is best regarded as an inexact popular term for the pupal stage if brilliantly coloured.

The estimated numbers of species in the text and in the appendix do not always agree, because they came from different sources. No one really knows just how many species exist in any insect group; new ones are being discovered all the time, while more detailed study shows that some of the existing names are synonyms, i.e. two or more different names really apply to the same species. About 8,000 new species of insects are discovered every year, and so figures like these are always out of date. Their value is to show the comparative sizes of different groups of insects.

The following corrections need to be made in the text:

Acknowledgements and page 2: plates 49, 50 were taken by Dr. H. Bänziger.

p. 19, and in Bibliography, for Demerecq read Demerec

p. 21, gall-midges do not dispense with adults *permanently*, but only in certain generations.

p. 45, for *Blatella* read *Blattella.*

p. 51, for Plecoptera read Psocoptera.

p. 53, for Trichodelidae read Trichodectidae, and for Mallophora read Mallophaga.

p. 56, figure 28: the bug should have a jointed proboscis like that in figure 53, p. 115.

p. 78, for *triangulin* read *triungulin.*

p. 115, for *Eugeron* read *Eugereon.*

plate 23, the larva is a lacewing (Neuroptera: Chrysopidae, cf. plate 24), not a ladybird.

p. 168, for Hippohoscidae read Hippoboscidae.

p. 190, for *Chalcidroma* read *Chalicodroma.*

p. 206, for *Ptenolepis* read *Prenolepis.*

p. 266, for *Metasequoia* read *Sequoia.*

p. 268, for *pseudospretena* read *pseudospretella.*

p. 101, item (a) (ii) should be moved to section (b), and on p. 157, the derivation of *Myrmeleon* is from *Myrmex,* ant and *Leo,* lion.

In conclusion, I hope that this list will not deter you from persevering with my book, and that you will find in it at least one or two ideas about insects that you can profitably pursue, through the Bibliography, and through direct observation on the living insects.

HAROLD OLDROYD

Introduction

The purpose of *Elements of Entomology* is to bridge the gap between school and university, or between courses of general biology and courses of zoology or specialist entomology. It is not a teaching manual but a book for background reading, in addition to recommended textbooks.

Insects are such a large class of animals that any university syllabus must be something of a crash-course. For examination purposes it is necessary to set questions that can be marked as right or wrong, or at least given marks on an arbitrary scale. Textbooks must therefore be packed with facts. Students learn these in sequence, and there is little time to gather these miscellaneous scraps of knowledge into a coherent picture.

On the other hand, the natural history of insects, outside the textbooks, is frequently presented as an assembly of disconnected facts. Yet the insects are a major group of animals, and one in active evolution. They have already shown themselves fully able to adapt themselves quickly to changes in the environment brought about by man. To become a working entomologist in any aspect of this science it is necessary to build up for oneself a coherent picture of insect life.

This book tries to encourage such an attitude. There is an irreducible minimum of names of parts, and facts of physiology, which must be mentioned, perhaps more than once, if they affect several aspects of insect life. As far as possible factual information is given in illustrations. The text is concerned with ideas about insect life, relationships and evolution. It is not suggested that this is a complete picture of entomology, but it is hoped that the book will suggest a number of different ways of looking at insects, and provide a framework into which facts from textbooks can be assimilated.

Such a book is a mixture of facts and ideas, the author's and other people's. A bibliography is given of works that are sufficiently comprehensive to serve as further reading, and works appearing in this bibliography are mentioned in the text with a number in square brackets. A problem arises, however, in respect of ideas and suggestions which are the work of a single author. His name can scarcely be omitted, yet to add all such works to the bibliography would make it too long, and too much like a textbook. So a few authors are mentioned in the text, and not in the bibliography, and their papers can be traced through textbooks and such works as *Biological Abstracts* and the *Review of Applied Entomology*.

I am very grateful to Professor Sir Vincent Wigglesworth, F.R.S., for advising me about the general content of the book, and to Mr J.F. Shillito for allowing me to pick his brains in matters of detail. The line-drawings owe much to previously published figures, to which I hope I have paid due acknowledgement: in particular to the textbooks of H.H.Ross and of Borror and Delong, both of which are invaluable to the serious student of entomology. The final drawings were made by Design Practitioners Ltd.

I am indebted to Mr Frank W. Lane for supplying photographs taken by the following photographers: G. R. Austing, 20; G. Baumann, 17; R. A. Brabbins, 10; L. W. Brownell, 9; R. L. Cassell, 13, 18, 21, 29, 30, 31; Lynwood M. Chace, 36, 38; T. Davidson, 22, 23; H. Doering, 14, 35, 46, 47; R. C. Hermes, 33; J. M. Hood, 44; M. Mihara, 48; W. Rohdich, 1, 15, 32; Leonard Lee Rue, 3, 34; F. Siedel, 8, 25; Hugh Spencer, 43; Edwin Way Teale, 39, 45; J. J. Ward, 19; Fred H. Wylie, 37. Dr W. Bütticker, who has made a special study of moths feeding from eyes, presented me with the two pictures by Dr H. Bäufiger (49, 50) and Dr A. de Barros Machado sent me plate 16 as a Christmas card. Plates 2, 3, 4, 6, 7, 11, 12, 24, 26, 27, 28, 40, 41, 42, were taken by myself.

As always, my wife has given me invaluable help and encouragement at all stages of the work.

1
Insects and other animals

Insects belong to the phylum *Arthropoda*, one of the major divisions of the animal kingdom. Arthropods are *invertebrates*, animals in which the body is not supported by an internal, articulated skeleton of bone or cartilage, but retains its shape by the rigidity of its outer surface. Arthropods are also *segmented animals*, in which the body is organised on a unit plan of basically similar *segments* (plate 1). This type of construction is not confined to arthropods: segmentation plays some part in the organisation of even vertebrate animals, and is fully developed in such animals as the annelid worms, which are not arthropods. In arthropods, however, the segmentation is emphasised by the hardening of large areas of the body-surface or *cuticle* into plates called *sclerites*. The sclerites of an arthropod form an *exoskeleton*, which is manipulated by means of individual muscles in place of the muscular layers of many other invertebrates.

The exoskeleton is all that is normally visible of an arthropod and gives it its characteristic appearance. The basic plan is that of an elongated creature something like a centipede, made up of a number of similar segments, each of which bears a pair of limbs or other appendages, which are themselves jointed: hence the name arthropods, the creatures with articulated feet (figure 2). Variations from this basic plan account for the tremendous variety of shape and structure found among the arthropoda, and are brought about by modifications of some or all of the segments and of their appendages.

Arthropoda are divided into a number of classes, each of which, in addition to internal differences, has a characteristic grouping of the segments into different functional regions, with associated modifications of the segmental appendages. Table 1 summarises these differences between classes. It will be seen that insects are characterised by the grouping of their body-segments into three regions: *head* (6 segments), *thorax* (3 segments), and *abdomen* (11 segments). This grouping is the only feature that is entirely diagnostic of insects; every other notable feature of insects is either subject to exceptions, or is shared with some other class.

Of these groups, only insects have wings, but wings are fre-

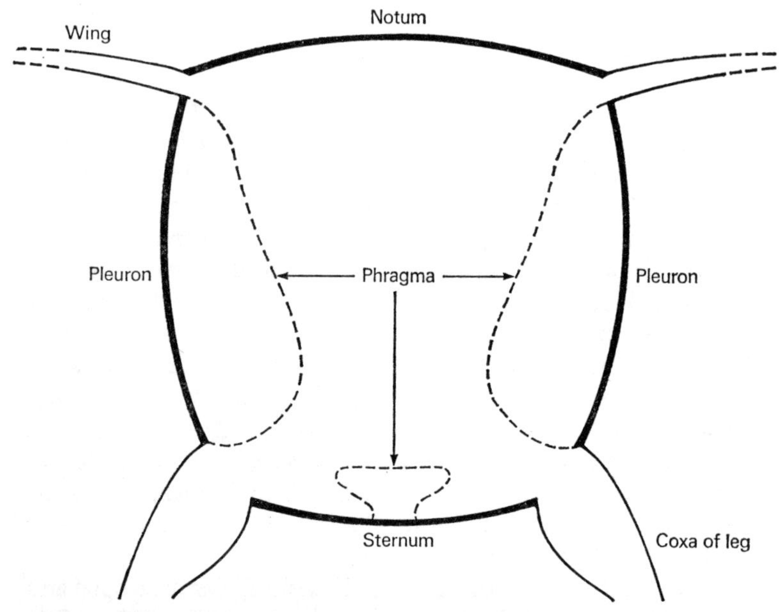

Figure 1. Component parts of an arthropod segment, shown diagrammatically.

quently lacking even in adult insects, and always in immature insects. An alternative name for the class Insecta is Hexapoda, referring to the three pairs of thoracic legs (figure 3). Only a few primitive insects retain traces of the appendages of the abdominal segments: e.g. *Thermobia*, the Fire Brat, shown in plate 2. In nearly all insects the abdominal limbs have been lost, except for those near the tip of the abdomen which form part of the external copulatory apparatus, and of the female ovipositor.

Similary, insects normally have one pair of *antennae*, or feelers on the head, and three pairs of appendages modified into *mouth-*

parts, but some of these may be modified or absent. Insects breathe air, conveying it directly to the tissues by means of a system of branching tubes or *tracheae*, but so also do the Chilopoda and Diplopoda (centipedes and millipedes).

Any generalisation about insects must specify whether or not it

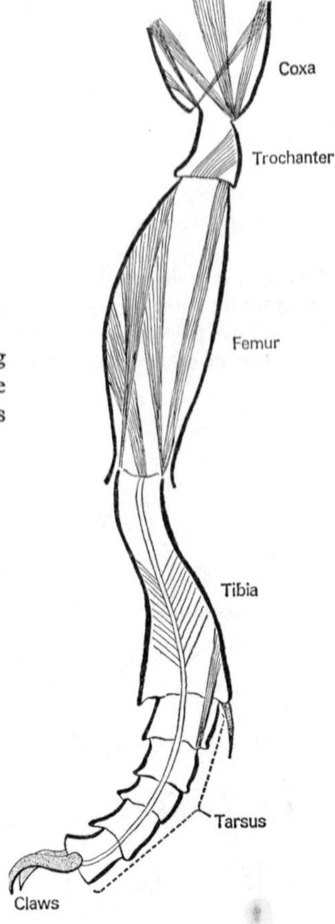

Figure 2. Segmentation and muscles of the leg of an insect. Note that the sub-divisions of the tarsus (tarsomeres) have no individual muscles (after Borror & Delong).

is related only to *adult* insects. Chapter 2 discusses the immature stages of insects, which sometimes closely resemble the adult but often are completely different. The arthropod form of construction, with its exoskeleton, does not entirely prohibit growth, especially in immature insects, but is particularly associated with a

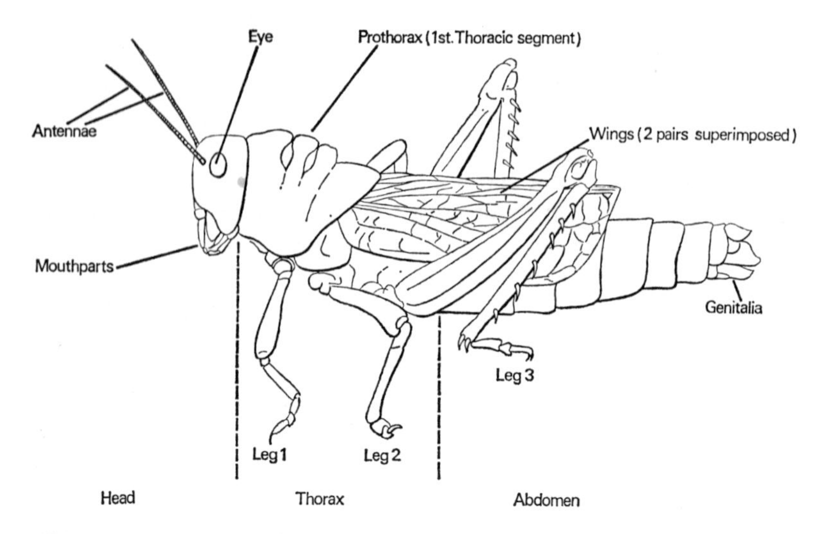

Figure 3. A male grasshopper in side view, showing grouping of segments into: Head – with antennae, eyes and mouthparts; thorax – with three pairs of thoracic legs and two pairs of wings (superimposed in this figure); and abdomen – without legs or other appendages, except for genital segments (after Klots).

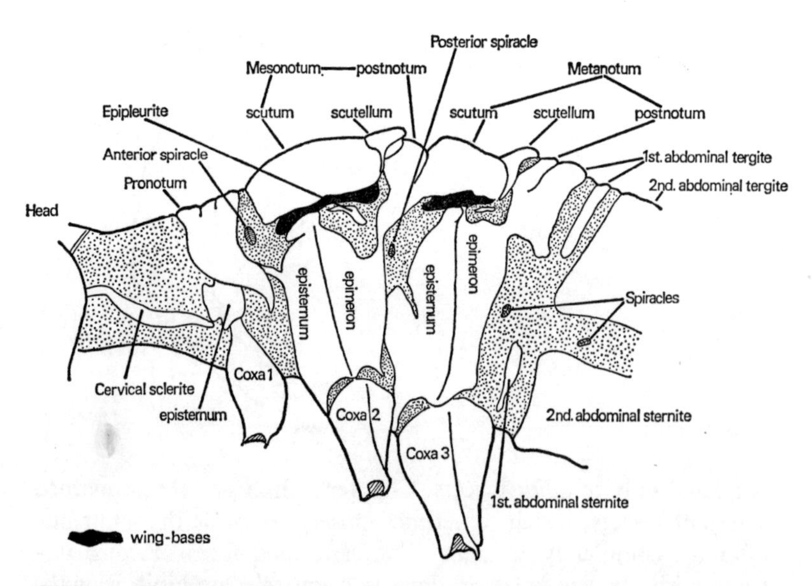

Figure 4. Thorax of the scorpion-fly *Panorpa*, showing sclerites (after Borror & Delong.)

discontinuous development, punctuated by a series of *moults* (*ecdyses*), when the outer layers of the cuticle are shed and replaced by a new growth. In any discussion of insect structure, therefore, it is essential to specify what stage of development is involved.

Comparative biology of arthropod classes

The four major classes of living arthropods are those listed in Table 1: Myriapoda (*centipedes:* Chilopoda and *millipedes:*

TABLE 1

COMPARISON OF SEGMENTATION IN FOUR MAJOR CLASSES OF ARTHROPODS

		Chilopoda		*Crustacea*		*Arachnida*		*Insecta*
1	HEAD	=====	HEAD	=====	PROSOMA OR CEPHALOTHORAX	=====	HEAD	=====
2		antennae		antennules		chelicerae		antennae
3		=====		antennae		pedipalpi		=====
4		mandibles		mandibles		1st legs		mandibles
5		1st maxillae		maxillules		2nd legs		maxillae
6		2nd maxillae		maxillae		3rd legs		labium*
7	TRUNK	maxillipeds*	TRUNK	maxillipeds*		4th legs	THORAX	1st legs
8		1st pair legs		thoracic limb		=====		2nd legs
9				thoracic limb	OPISTHOSOMA OR ABDOMEN	genital ♂ openings ♀		3rd legs
10				thoracic limb		pectines*	ABDOMEN	
11				thoracic limb		1st lung-books		
12		22 segments with genital opening posteriorly		♀ opening		2nd l.b.		11 abdominal segments with genital opening posteriorly
13				thoracic limb		3rd l.b.		
14				♂ opening		4th l.b.		
15						no limbs		
16				6 abdominal limbs				
						5 segments of metasoma		

======= denotes a segment that can be traced during embryonic development, but which cannot be distinguished later.

* *pectines* are sensory organs found in scorpions and their relatives; *maxillipeds* are supplementary mouthparts modified from the first pair of walking limbs; the *labium* of insects is a single structure formed from the union of a pair of appendages, equivalent to second maxillae.

Diplopoda); Crustacea (crabs, lobsters and many related aquatic forms); Arachnida (spiders, scorpions and mites) and Insecta (insects).

Myriapoda have progressed little from the ancestral condition, and, though sometimes numerous as individuals, they are an insignificant part of the arthropod fauna of today. The other three classes named have all been highly successful in evolution, but in different ways.

Crustacea are predominantly an aquatic group, and particularly marine. They have made much use of the characteristic jointed appendages of the arthropod form of construction, and use their limbs for a variety of purposes, for swimming, for moving along the sea-bottom, for feeding, for tactile and other senses, and, with modifications, even as respiratory organs.

As befits an aquatic group, nearly all Crustacea breathe the oxygen dissolved in water. Tiny Crustacea can get as much oxygen as they need through the general surface of the body; larger ones possess *gills*, specially enlarged areas of thin cuticle modified from the bases of some of the legs. Some of the terrestrial Crustacea, such as the land-crabs, use virtually the same mechanism as their aquatic relatives, merely keeping the gill-chamber full of air, but the truly terrestrial Crustacea belonging to the order Isopoda, the woodlice, have developed tracheae similar in function to the tracheae of insects, though quite independently evolved.

The other three classes of arthropods are ecologically the reverse of Crustacea, in that they are essentially terrestrial groups, in which a few members have returned to life in water. Arachnida are a strange group, of great diversity, ranging in size from the giant king-crab, *Limulus* down to microscopic, colourless and almost invisible mites. Among living arachnids, only the king-crabs are aquatic, and they live on the seashore, burrowing in the sand. Respiration in arachnids is by a number of different organs; gill-books in *Limulus*; lung-books in scorpions; tracheae in mites and pseudoscorpions; and both lung-books and tracheae in spiders.

Arachnida, therefore, are not only an evolutionary line quite widely divergent from insects, but they show a greater variety of basic structure than do the insects. Some of them, notably spiders and mites, become involved in the problems of practical entomology in relation to crop plants and to disease, and from time to time there is the suggestion that 'entomology' ought to be broadened to include spiders and mites; but there is no zoological warrant for this.

The class Myriapoda is of interest to the entomologist chiefly for the light that it throws on the probable origin and evolution of insects. A common centipede from the garden comes close to the idea of a primitive terrestrial arthropod. The body is made up of a series of similar segments, each with its pair of jointed appendages, here modified, as befits a terrestrial animal, into walking legs. Only at the two extremities of the body are these limbs used for other purposes, at the head as antennae and posteriorly as genital organs. Air is breathed by means of tracheae.

Characteristics of insects

Insects as a group are land-living and air-breathing, though many modern insects now live in the water for at least a part of their life. Those insects that have come back to live in the water, or in any other medium in which atmospheric oxygen is inaccessible, notably as parasites in the tissues of other animals, have had to adapt their tracheal system, rather clumsily, to do something for which it was not originally evolved. To deal with temporary immersions in water insects make much use of a *plastron*, or thin film of air within which an exchange of gases can take place; but insects that are truly aquatic have had to close their tracheal system and, if they are to get enough oxygen for active life under water, have had to expand the tracheal system into tracheal gills (see Chapter 12).

The legs of insects are also obviously devices suited to life on land. The superficial resemblance between the leg of a cockroach (figure 2) and the leg of a vertebrate is an example of evolutionary convergence. Although the two kinds of limb have a fundamentally different structure, the insect's leg being a hollow shell with the muscles inside, each kind of leg is a similar mechanical device which raises the body of the animal from the ground and then propels it forward by pushing against a rigid surface.

Again, it is true that some insects have come to live in water and have modified their legs – or some of them – into oars or paddles. Figure 46 shows an example of such modifications for aquatic life. These, however, are secondary modifications, which affect only parts of the legs of a tiny minority of insects. The very fact that a similar practical result has been obtained in different ways in many different groups of insects only serves to emphasise that insects as a class are a terrestrial group, and a relatively uniform one at that, compared with Arachnida.

It is thus fairly safe to assume that insects arose from some ancestral arthropod that had already left the water and was fully terrestrial. This ancestor must have been at least as primitive as the most primitive living insects. These are considered to be the bristle-tails of the order Thysanura (see page 30 and plate 2). They have, of course, the segments grouped into head, thorax and abdomen, but the segments of the abdomen may still retain their simple appendages, which more advanced insects have lost. Their nearest living relatives outside the insects are the Symphyla, a group which comes between the centipedes and the millipedes. They look like primitive insects, and the resemblance is greater still if we compare the immature stages.

A major difference between the adults of millipedes and insects lies in the position of the external genital opening. As shown in Table 1, Chilopods and insects have this opening posteriorly, whereas Myriapods, like Crustacea have the opening on a segment which, in insects, would correspond to the beginning of the abdomen. Myriapods are *progoneate*, insects are *opisthogoneate*. The internal genital organs, or gonads, like so many Arthropod structures, were originally segmental, and became linked together to form a compound organ. The position of the external opening has been interpreted by many authors as an indication of a basic difference in internal organisation, important enough to prevent the insects from being regarded as descendants of primitive Myriapods.

Other authors, however, dispute this view, and think that the position of the genital opening is a secondary development. Manton (1953) gives a fascinating account of locomotion in various groups of Arthropods, and comes to the conclusion that: 'The gaits of the Symphyla, alone among Myriapoda, basically resemble those of Insecta and could readily give rise to the latter by further advancement of the locomotory mechanism.'

There is therefore good authority for accepting the view that insects arose from an ancestor among Myriapods, and that the nearest relatives of this ancestor that are still in existence today are the primitively wingless Apterygota (page 30).

2
Immature insects

Vertebrate animals become adult gradually, as a continuous process, which may continue after sexual maturity has been attained. There is no precise moment at which a vertebrate body as a whole becomes mature. In contrast to this the development of an insect is a discontinuous process, punctuated by a series of *moults*, at which the cuticle is shed, and the insect emerges with a new cuticle, usually differing in some respects from its predecessor, and sometimes profoundly changed. Since the insect's relations with the outside world are dependent almost entirely upon the organs of the cuticle – organs of sight, hearing, touch, smell and taste, and powers of locomotion – a change of cuticle at a moult may be linked with sudden changes in the behaviour and biology of the insect. A familiar example is the moult from a caterpillar to a pupa in its chrysalis, followed by another moult to the adult butterfly.

A moult does not necessarily bring changes as extreme as these. In the most primitive insects, which never have wings, moulting may continue throughout life, with no obvious change in external appearance. Nor does an insect necessarily grow bigger after a moult. Often the newly formed cuticle, soft and much folded, is rapidly inflated before it hardens, but primitive insects, with many moults, may actually become smaller if they have been starved of food. One consistent function of moulting is to eliminate certain waste products of bodily metabolism.

The wings of insects are an obstacle to moulting. Figure 7 shows how the wings develop from the pad-like outgrowths of the thorax, which can be moulted as long as they remain pad-like, but which only become functional as wings after they have become compressed into a membrane supported by ribs. This structure cannot be moulted effectively, though in one group of insects – May-flies of the order Ephemeroptera, a primitive order of winged insects (Chapter 3) a moult may occur after the wings have expanded. The fact that no other winged insects moult after expanding the wings suggests that this obstacle to further moulting has proved effective during the evolution of insects.

Insects technically become adult at the moult after which the internal sexual glands become functional. In winged insects this is nearly always the moult at which the wings are expanded, and after this no important bodily changes occur, though in certain groups that have been specially studied – for example certain disease-carrying flies, where some method of estimating the age of individual adult flies may be helpful in studying the history of the disease – it has been possible to detect small changes in pigmentation that go on throughout adult life.

Stages of Insect life

The life of an insect falls into three stages: the *egg*, during which it is an *embryo*, an organism not yet capable of an independent existence; a period of immaturity extending over a number of moults; and the period after the final moult (in winged insects) or after sexual maturity (in primitively wingless insects), when it is an adult insect, or *imago*.

The expression 'period of immaturity' is misleading if it is understood in the sense of childhood, or of the young of vertebrate animals or birds. Social insects such as bees or ants have an immaturity during which they are helpless and have to be looked after by others of the community, but among insects generally the so-called 'immature' phase of life is self-supporting and self-sufficient, seldom with any contact with its parents. Its contribution to the evolutionary progress of the species may be as great as, or even greater than that of the adult. The caterpillar of a moth, or the maggot of a fly has a life of its own, and should not be thought of merely as 'waiting to grow up'.

Except for the sterile workers of social insects, whose business in life is to tend the young, the particular function of adult insects is sexual reproduction. It seems reasonable that mating should be deferred until after the final moult in winged insects, because the insect then becomes much more mobile, allowing both for the spread of the species and for the mixed matings which produce variety and

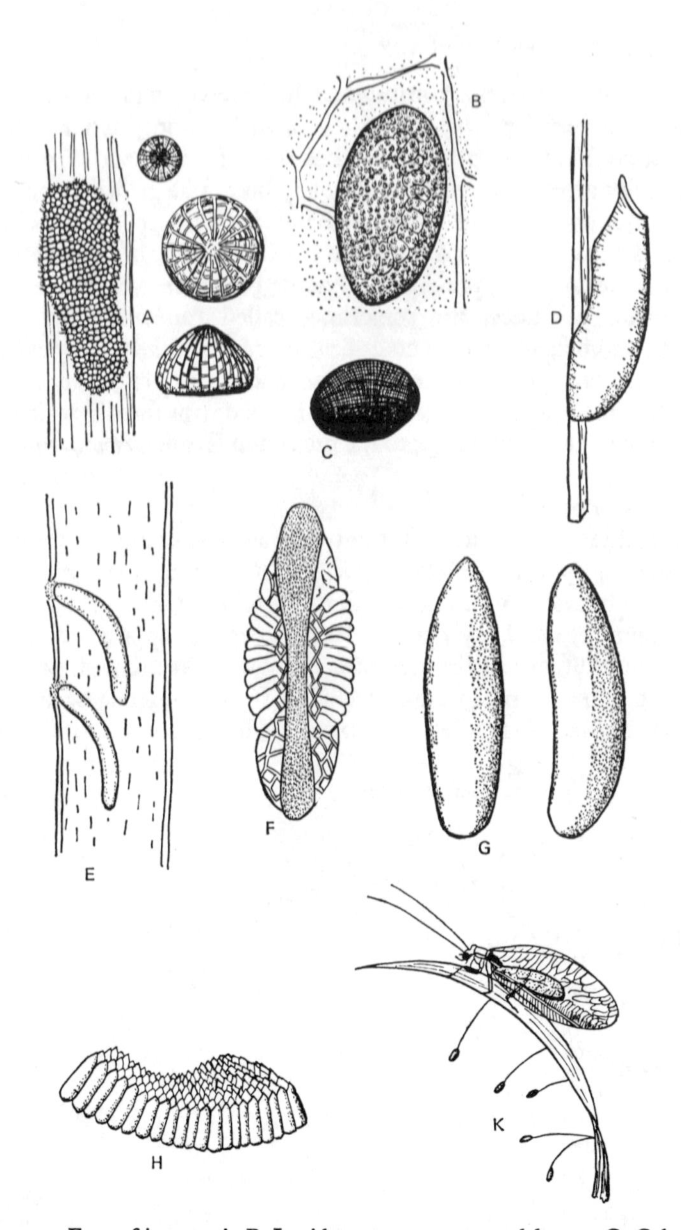

Figure 5. Eggs of insects. A, B, Lepidoptera, on stems and leaves; C, Coleoptera, on ground; D, Diptera, attached to a hair (horse bot-fly); E, Orthoptera, a tree-cricket, inserted into a stem; F, Diptera, an anopheline mosquito, egg with a flotation device; G, Diptera, in soil; H, Diptera, culicine mosquito in a floating raft; K, Neuroptera, Lacewing fly, stalked egg attached to a leaf (after Borror & Delong).

evolutionary progress. Until insects had evolved wings they could not move very far away from their birthplace, only as far as they could run. Life after hatching consisted of a number of moults, generally growing steadily, and finally becoming sexually mature, mating and laying eggs, and then dying. Such insects changed little in outward appearance throughout their life. Some of their descendants live today as the Apterygota, or primitive wingless insects (Chapter 3). These are sometimes called Ametabola, meaning 'without change'. All other insects, including many wingless insects, that are considered to have descended from winged ancestors, are Metabola, and the series of transitions from the form in which the insect hatches from its egg to the final adult is called *metamorphosis*.

Metamorphosis

Winged insects are divided into two groups according to the type of metamorphosis they undergo. The more primitive group, i.e. the one believed to have come first in evolution, has a progressive metamorphosis. They emerge from the egg in appearance much like the adult, except for the absence of wings. At successive moults they become even more like the adult by small changes, and the wings appear as small external pads. At the final moult the wing-

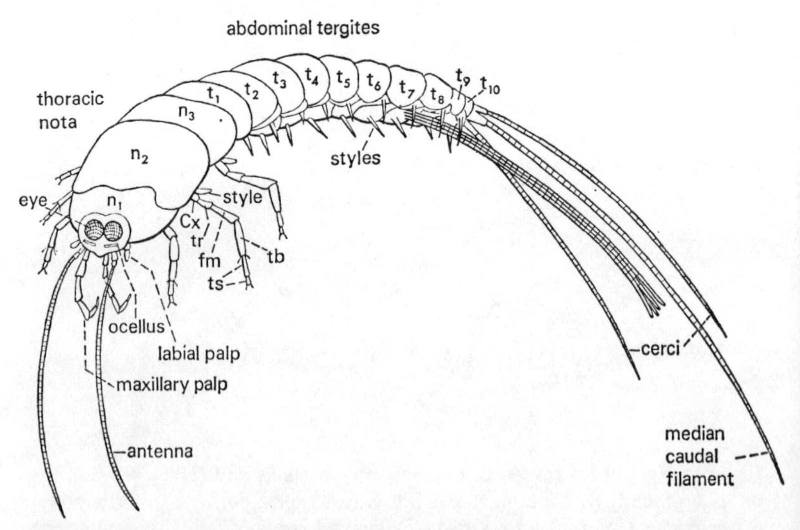

Figure 6. A silverfish (order Thysanura), a primitively wingless insect (after Borror & Delong) (see also plate 2).

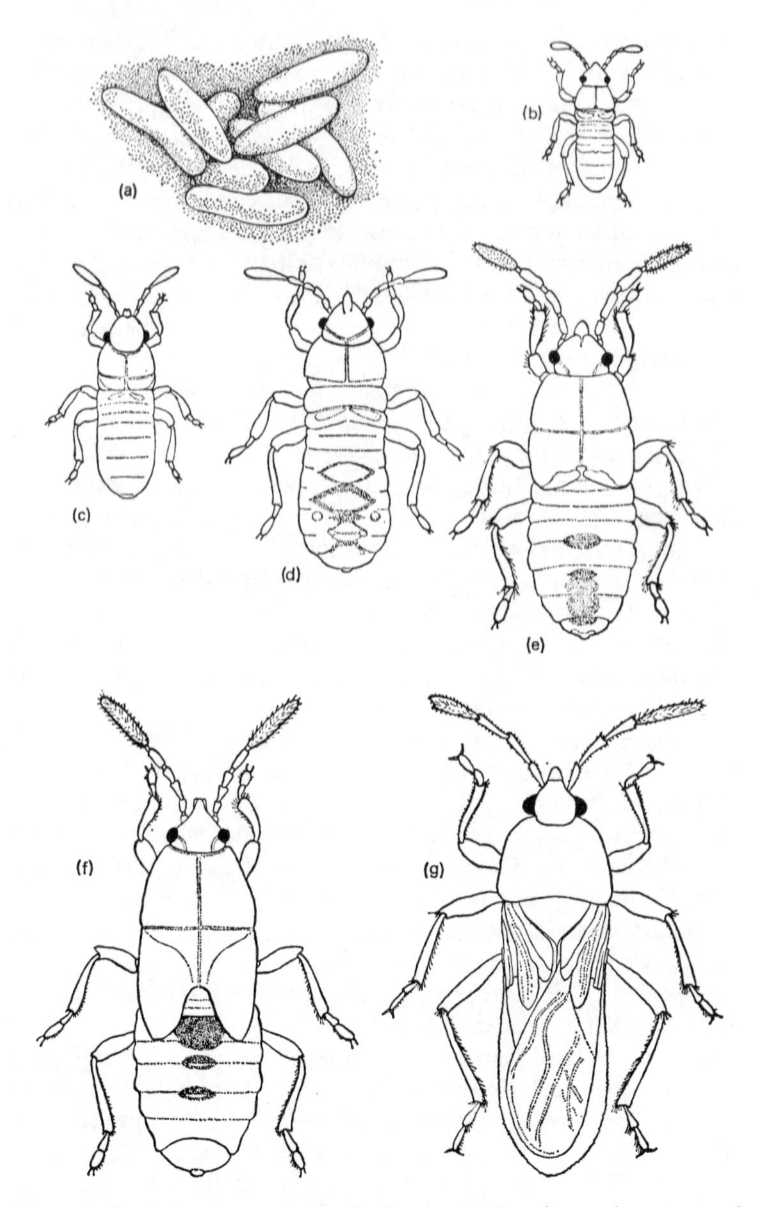

Figure 7. Gradual metamorphosis of a hemimetabolous insect: 'exopterygote' because the wings appear externally as wing-pads, but only become operative after the final moult (after Borror & Delong).

pads expand into functional wings (figure 7). This group of insects is known as *Exopterygota*, because the wing-pads develop externally, and *Hemimetabola* because their metamorphosis is, as it were, only a half-hearted affair. Into this sub-class come such insects as cockroaches, locusts and grasshoppers, Dragon-flies and May-flies, and aphids and other plant-bugs. (see Chapters 4 & 5) Some of these insects spend their immature stages in water, and then the exigencies of aquatic life have led to the development of special devices for breathing and moving under water (see Chapter 12). To this extent they depart from the rule that the immature stages of *Hemimetabola* resemble the adults.

Other insects of later evolution have what might be termed a more positive approach to metamorphosis. Instead of leading gradually up to the adult they have exaggerated the differences between immature and adult phases. The caterpillar of a moth, or the maggot of a fly does not in the least resemble its parents, or its own adult self, and what is more, it does not become any more like them as it grows older. It has become an independent creature with a life of its own, and is called a *larva*. (Some modern authorities use the term larva for the immature stages of all insects but it is traditional, and convenient, in English at least, to call the immature stages of the Hemimetabola *nymphs*.)

A succession of immature stages which do not progress towards the adult has one obvious drawback: sooner or later the insect must become adult. For a nymph this is a short step, just a final unfolding of the wings, but a larva has to change into what is virtually a different creature. The metamorphosis of the caterpillar into the butterfly has become the proverbial example of a spectacular transformation. That is why this group of insects are called *Holometabola*, those having a complete metamorphosis. They are also called *Endopterygota* because the wing rudiments do not appear as external pads, but can be traced internally as wing-buds.

The gap between larva and adult is bridged by a 'resting' stage known as the *pupa*. 'Resting' is a misnomer, since in fact this is one of the most profoundly disturbed periods of the insect's life. Though the outward appearance is that of a still, mummy-like creature, inside the cuticle all the tissues of the larva are being broken down and remodelled into those of the adult.

The various stages of the life-history of an insect may now be considered in turn.

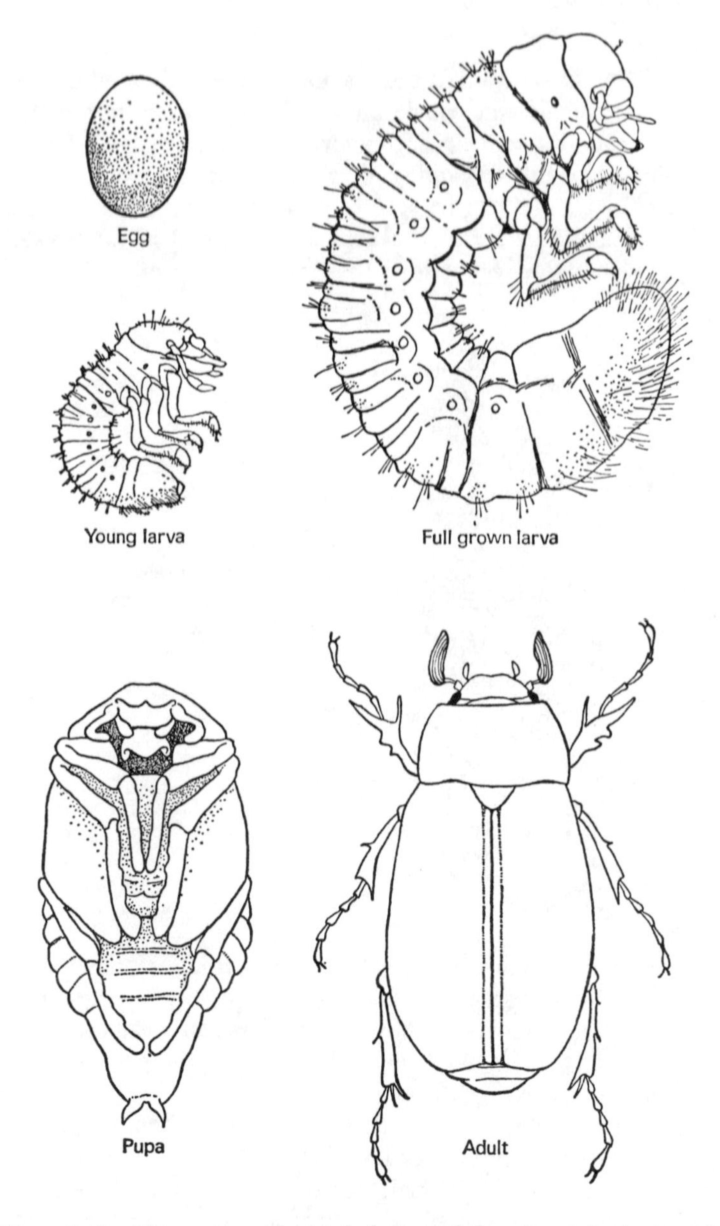

Figure 8. Complete metamorphosis of a holometabolous insect: 'endopterygote' because no trace of wings is visible externally until the pupal stage is reached (after Ross).

The Egg

Insects begin life as a single cell, its nucleus normally produced by the fusion of male and female nuclei. The cytoplasm of the egg, which surrounds the nucleus, contains yolk or food material for the nourishment of the developing insect until it is ready to hatch.

Normally all that can be seen of an insect's egg is its external surface. Many eggs are ornamented or sculptured in various degrees of elaboration. Sometimes there are stalks, or other forms of

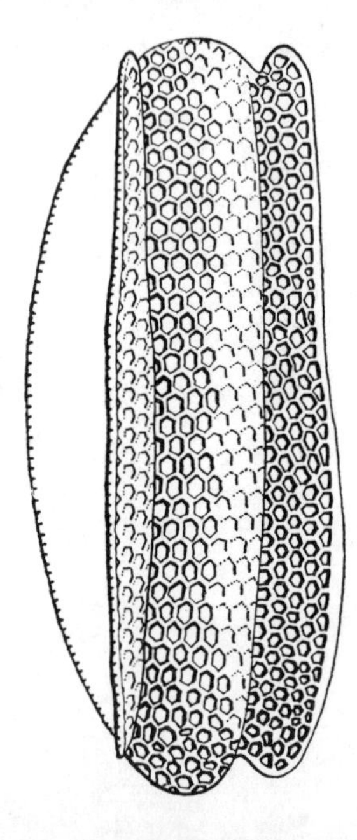

Figure 9. Egg of a fly, showing reticulated outer surface, associated with retention of a plastron when submerged by rain.

attachment by which the egg may be stuck to a leaf, or to some other surface, and eggs that are laid in water may have some sort of flotation device. Most often, however, the structural elaboration of the surface of the egg has to do with regulating the passage of air and water in and out of the egg.

Most insects lay their eggs – that is, they push them out of the oviduct (egg-tube) and either let them fall to the ground, or into water, or place them individually in position on or beneath the surface of a plant. Only a small minority of insects give their eggs any further attention, and all eggs that are exposed to the weather have to be able to resist violent changes of heat and cold, flooding and drought. The embryo within the egg must respire, so there must be pores through which oxygen can enter, yet water must not penetrate beyond the amount that the insect needs. Hinton's paper [44] shows how eggs submerged in water can continue to breathe by the use of a plastron, or thin air-film similar to that used by certain adult insects that live in water (see Chapter 12).

So the elaborate surface ornament of many eggs is more than mere decoration, and has important functions of its own. Besides resisting the drowning of eggs that are submerged, and the desiccation of those laid in exposed situations, the surface of some eggs can absorb water from their surroundings. Eggs of some parasitic insects also absorb nutritive substances from the surrounding tissues of their host animal, to supplement their own inadequate supply of yolk.

The single cell of the egg divides again and again, and builds up the various layers of cellular tissue in a way similar to that studied in the embryos of vertebrates. The embryology of insects is a complex field of study: a useful introduction to the subject is the account of the embryology of the Fruit-fly *Drosophila* given by Demerecq [29].

Usually this development from a single-celled stage does not begin until the egg has been fertilised, but this is by no means universal. Parthenogenesis, or the development of an embryo from an unfertilised egg, is widespread among different groups of insects, of which the best known are the social insects, bees, wasps and ants, where unfertilised eggs give rise to females, and fertilised eggs to males. This device, if one can call it that, has arisen independently many times. It seems as if the possibility is always latent, and may be called into action whenever it is an advantage to the insect, when natural selection causes a parthenogenetic strain of the insect to develop. The obvious biological advantage is that parthenogenesis speeds up reproduction, and consequently the increase in numbers of a population. Sexual reproduction requires times for the two sexes to find each other, and to carry out courtship behaviour, mating flights and so on. On the other hand there is none of the mixing of hereditary material that may result from sexual reproduction, especially if the adult insects have considerable powers of flight.

A few insects seem to find that the general advantages of parthenogenesis outweigh the disadvantages: for example, the stick insects,

(Phasmida) among which males are rare. These insects remain motionless for long periods, camouflaged against their enemies, and any movement may give them away to an enemy. Dispensing with the need to look for a mate may be seen as part of their negative approach to the problem of survival.

Other insects use parthenogenesis in a more positive way. Aphids (Green-fly: Black-fly) and some gall wasps (Hymenoptera: Cynipoidea) may alternate between periods of parthenogenesis and sexual reproduction. The former enables them to increase in numbers almost explosively during the summer months, while sexual reproduction during the winter months mixes up the hereditary material and physically distributes the insects over a wider area.

The fact that unfertilised eggs may produce males in some insects and females in others is evidence that the phenomenon of parthenogenesis has risen more than once during the evolution of insects.

Embryonic development within the egg is usually postponed until the egg has been laid, and sometimes for very long after that. Eggs of certain stick insects may lie on the ground for nearly two years before hatching. The normal period in favourable circumstances is from a few hours to a week or two, but the egg stage is often used as a 'buffer' period against unfavourable conditions. In temperate countries the winter may be passed as an egg: in hot, arid conditions the time spent withdrawn inside the egg may be a convenient way of surviving the dry season. Such a period of inertia, which carries the insect through an unfavourable season, or temporary inclement conditions, and which requires a definite stimulus to bring the inertia to an end, is called a *diapause*.

At the other extreme, some insects seem in a hurry to get into the world, and the period during which the egg exists away from its parent may be curtailed or even abolished, as in the so-called 'viviparous' insects, though this is an inaccurate term, since all insects have a true egg stage, however short. The appearance of viviparity may come about in several different ways. The simplest is that the eggs just hatch prematurely, before they are dropped by the mother insect. The next stage is that the larva is provided with food from special glands in the 'uterus' of the female (these terms such as 'uterus' are placed in inverted commas because they are taken by analogy from the biology and anatomy of vertebrate animals, and do not properly suit the different biology of insects). An example of this type is the Tsetse-fly *Glossina*, which nurtures its larva until this is fully grown. The larva feeds normally through its mouth, not through a placenta like a mammal.

A further development is when the larva receives nourishment, not through its mouth, but through its skin, and then more nearly resembles the embryonic mammal in its womb. A final grisly stage is reached when there are no oviducts, and no provision at all for laying the eggs, which lie scattered through the body-cavity of the mother insect, and there hatch into larvae, which eat their way out of the mother's body when they are ready to leave. This happens in some gall-midges (Diptera: Cecidomyiidae), which, with a grim evolutionary logic, dispense with the adult stage altogether. A larva produces eggs, and then larvae, within its own body-cavity, and is ultimately devoured by them. This way of reproduction from an immature stage, without the intervention of an adult, is called *paedogenesis*.

Nymphs and Larvae

At the end of the egg-stage comes the first moult, one of a number during the life of the growing insect. Moulting, or *ecdysis*, takes place in two stages. First, the insect detaches itself from its own cuticle, and grows a new cuticle inside the old, at the same time making any necessary changes in structure appropriate to the new *instar*.* Then, as a separate step, the insect breaks out of its old cuticle and discards it. The term 'moulting' is usually applied to the second step of breaking out of the old cuticle, because this is easily observed, but the first step, the actual transformation of the insect, is the important one. Usually there is only a short interval between the two events, and this lack of precision in terminology does not matter. Sometimes, however, there is a significant period during which the insect has already entered into the next phase, even though it is still confined within the old cuticle. Such an interval is called a *pharate*, or concealed stage.

This apparently trivial distinction has only recently been stressed. As an example, an insect that is known to survive the winter in the egg may do so in two different ways. The embryonic

* 'The intervals between the ecdyses are known as stages or *stadia*, and the form assumed by an insect during any particular stadium is termed an *instar*.' Imms [50].

development within the egg may slow down or cease in a winter diapause, to be resumed in the spring; or development may continue more or less normally until the embryo is ready to hatch, but the embryo may remain within the egg-shell until warmer weather comes. In the second event the insect has passed the winter, not as an egg, but as a *pharate nymph* or *pharate larva* as the case may be.

It is convenient, in English, to use the term *nymph* for the immature stages of Exopterygote insects, reserving *larva* for the immature instars of Endopterygote insects. Some English entomologists object to this practice, either because of possible confusion with the French *nymphe*, which means a pupa, or because they support some particular theory of the origin of metamorphosis, and think the distinction is misleading. While bearing in mind these points of

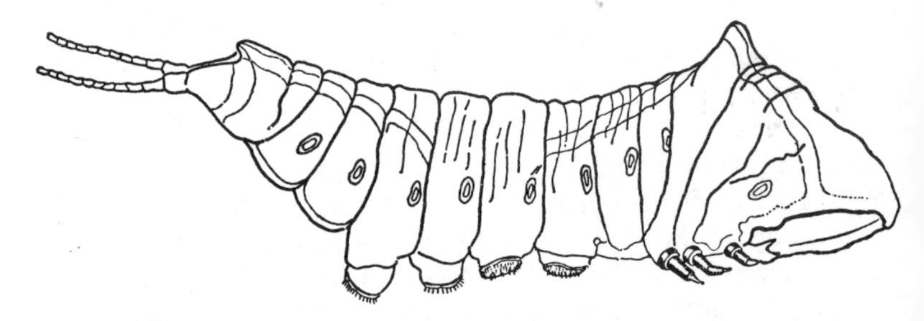

Figure 10. Caterpillar of a puss-moth, example of an eruciform larva.

view, it is still the commonest practice to use the terms as explained above. The name *naiad* is sometimes used for the aquatic nymphs of Dragon-flies and similar insects.

Nymphs resemble their parent insects only if they live the same kind of life: for example Green-fly on a stem, the nymphs feeding alongside the adults; or young grasshoppers (these, particularly young locusts are often called just 'hoppers') or plant-bugs feeding on vegetation like their elders. Quite a few Hemimetabola (= Exopterygota), however, have adopted a different way of life when young and when adult. This particularly applies to an aquatic life in ponds and streams. Quite early in the evolution of winged insects some nymphs took to water, and even today the nymphs of such well-known insects as Dragon-flies (Odonata), May-flies (Ephemeroptera), Stone-flies (Plecoptera) and Alder-flies (Neuroptera) live in water, and only leave it for the winged adult stage.

Life in water is an exacting one, which calls for specialised equipment for breathing, for swimming and for getting food. Aquatic nymphs must therefore possess the necessary equipment, which is discarded at the final moult if it is not required by the adult insect. The nature of the problem of life in water is essentially the same for all insects, whether they are adult or immature (see Chapter 12).

Larvae have been able to diverge entirely from the shape of the adult, because the intervening pupal stage makes it possible to metamorphose into adult shape at one big step. The caterpillars of butterflies and moths, the grubs of beetles, bees, wasps and ants, the maggots of flies, are all larvae. None of these looks in the least like its parents, and it is fruitless to try to see resemblances. Larvae should be looked upon as creatures in their own right, and considered in relation to the life that they lead during the larval stages of their development.

Larvae considered in relation to their habits and habitat fall into three main types, in order of decreasing activity. The most active are those larvae that have three pairs of jointed legs, a well-sclerotised head and thorax, and an elongate abdomen. They bear a marked resemblance to primitive wingless insects, in particular to the bristle-tail Campodea (order Thysanura), and therefore they are called *campodeiform* larvae (figure 14).

Since these larvae have evolved their particular structure in response to the demands of some particular environment, some correlation between the two might be expected. Such larvae are generally predaceous, often fierce and voracious. Examples are the larvae of Ground-beetles (Carabidae) and those of the various groups of Neuroptera, the Alder-flies, the Lace-wing-flies (plate 24) and the fierce ant-lions which Wheeler called 'Demons of the Dust'. The last deserve this melodramatic name because they lie at the bottom of a conical pit in fine sand or dust and feed on insects such as ants which fall into the pit.

A degree less active are the *eruciform* larvae, those which resemble caterpillars (figures 8 & 10). Caterpillars are grazing animals, and crop their way with tireless efficiency over a patch of vegetation, like the herd of six-legged sheep or cows. They have segmented walking legs on the three thoracic segments, but these legs are relatively short. The body as a whole is generally soft and inflated, and the sagging abdomen may be supported on a series of false legs: i.e. these are not true segmental appendages, and probably have no connection at all with the original segmental limbs of the abdomen, long lost in evolution. The false legs are pad-like swellings of the cuticle, often armed with rows of small hooks, or *crochets* which help the organ to grip the substratum. Such false legs are called *prolegs* or *pseudopods*.

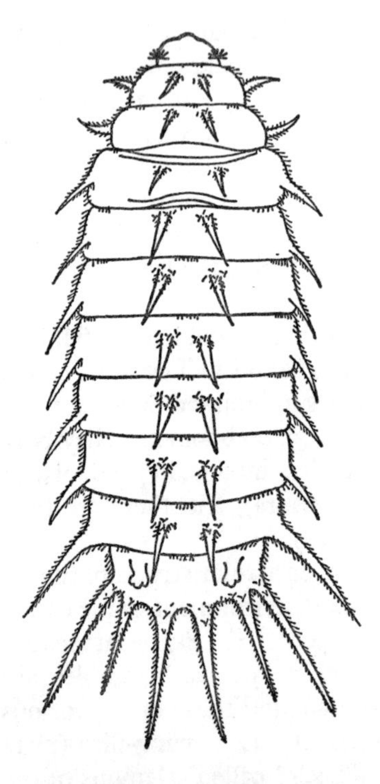

Figure 11. Larva of the fly *Fannia*: a maggot (cf. Figure 73 C), with secondary development of fringed processes, in adaptation to living in a very moist medium (after Oldroyd).

The abdominal segments certainly participate in movement, anchoring the prolegs in successive pairs, and then pulling the body forward against this resistance (figure 10). Those prolegs that face downwards are often the most highly developed, but prolegs may occur all round the abdomen, and may be used to push against any surface, e.g. to squeeze through a small hole, or to wriggle through the soil.

Eruciform larvae can certainly move, and sometimes very quickly, but the flabby body and cumbersome rhythms of abdominal movement condemn these larvae to crawl rather than run. They are thus typically plant feeders, though the eruciform larvae of the well-known ladybirds, beetles of the family Coccinellidae, 'graze' among the aphids and other plant-bugs that encrust many a stem (plates 22, 23). Films taken in close-up of the ladybird larvae in action show that this metaphor is a true one. It is a macabre spectacle to see a Green-fly being seized and eaten while its neighbours go on feeding without taking any notice.

Larvae that live *inside* their food material, instead of merely on it, rely even less on movement. The big, juicy stag-beetle grubs that can be found in a rotten stump have thoracic legs, but the abdomen is so grossly swollen and dropsical that the idea of their ever running is ludicrous. 'Obese' is the only word for them. Progressive reduction of the thoracic legs as they become superfluous leads to the third general type of larva, the completely legless or *apodous* type (figure 71). This state is attained by certain other beetles, so that within one order of insects can be found the whole range of larval adaptation. Apodous larvae are also characteristic of the bees and wasps, the ants and the parasitic Hymenoptera, insects whose larvae live an entirely protected life, either as a parasite inside the body of another animal, or in a waxen or paper cell where they are tended by adult insects. The larvae of all flies and fleas are also legless, even though those of many flies have returned to an extremely active way of life. For this they have had to evolve new structures, and all larvae seem to be able to do this, if the evolutionary pressure is great enough.

Flies are a particularly good example of this (figures 11 & 72). Apparently they ceased to have legs in the larval stage quite early in their evolution and have had to get along without them ever since. Yet as a group flies have colonised all the possible places where land animals can live, including all the aquatic habitats. There are larval flies that swim about in deep water, coming to the surface to breathe; some that attach themselves to submerged rocks and underwater plants; others that crawl about under water as if they were on land: and yet others that pierce the stems of aquatic plants and tap the supplies of air contained within them.

All these activities call for special equipment, and flies have proved fully equal to the challenge, evolving elaborate structures which help them under water, just as efficiently as the more primitive insects that we have already discussed. Among flies the evolutionary trend has been to simplify and streamline the larva. The larvae of the more primitive flies are more or less obviously adapted to their way of life; those of more highly evolved groups of flies are all maggots of one sort or another, a universal type that can adapt itself physiologically to life in almost any place. Yet even here the pressure of life in water, and of certain kinds of parasitic existence, has led some of these larvae to develop special equipment.

Pupae

Insects that have larvae also have pupae. The pupal stage is a mysterious part of the life-history, often dismissed briefly as a 'resting' stage between larva and adult. It has the appearance of an inert resting stage for two reasons: because the insect is generally motionless through most or all of this period, and because many insects take advantage of the pupal period to pass through an unfavourable period of the year. This may be the winter in a temperate country, or a hot, dry season in the tropics.

Confusion sometimes arises between the pupa and the pharate adult, the mature insect already fully formed, and lying inert within the pupal skin, awaiting its time to come out. Most of the movements attributed to pupae really refer to the pharate adult, yet it cannot be said categorically that all true pupae are inactive.

It appears that the primary function of the pupa is the mechanical one of remodelling the larval mechanisms into those of the adult. If some of the larval muscles are also needed by the adult, then these can continue to exist, and even to function, during the pupal stage. Hence the amount of movement that is possible to a pupa varies according to the insect concerned.

Once again, those insects whose pupae are aquatic have to provide them with a means of breathing, and of regulating the osmotic pressure within the body in relation to that of the surrounding water. It is mostly these pupae that are able to move, or even to swim, for example pupae of mosquitoes. Pupae that are buried in the soil, or in burrows in wood, are often equipped with powerful spines or tooth-like structures, which are used to force a way to the surface just before the adult emerges. Though the structures themselves may be part of the pupal cuticle, the effort behind them is often provided by the pharate adult, which does not emerge until the pupal skin has reached the open air.

A pupa is not a creature with a life of its own, as we have seen a larva to be, and consequently there is not properly a pupal shape.

Since the pupal cuticle serves as a mould for the shaping of the body of the adult, it has a simplified outline of the adult insect. The word 'pupa' means a baby or a doll, and this expresses the softened outline and lack of precise detail that we find in the pupae of insects.

Pupae may be classified by their external appearance into two groups: *exarate* pupae, in which the wings, legs and antennae are

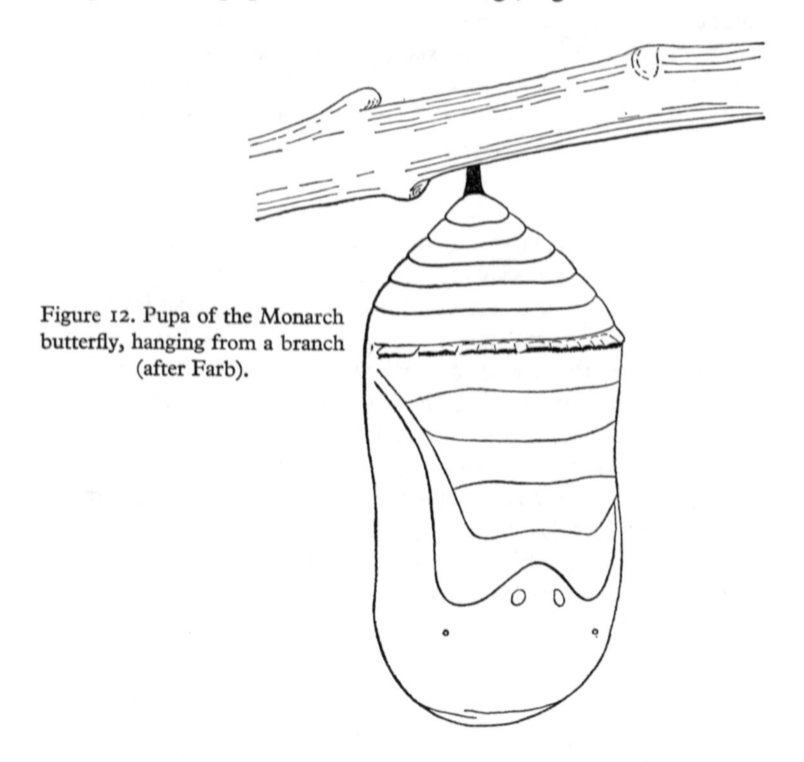

Figure 12. Pupa of the Monarch butterfly, hanging from a branch (after Farb).

all free, and so the pupa looks like a rather bedraggled adult (figure 8); and *obtect* pupae, where these appendages are stuck down to the surface, making the pupae look like an advanced piece of sculpture. The pupa is very often enclosed in an outer casing or *cocoon*, which is generally constructed by the larva from extraneous materials, soil, mud, sand or earth. Silk is a fibre used by many insects in the construction of their cocoons (e.g. silkworm), and of other larval shelters. Caterpillars of butterflies and moths and larvae of Caddis-flies produce silk from glands attached to the labium, the third pair of mouthparts; other insects produce it from glands in the tarsi of the legs, or even as a waste-product of excretion from the Malpighian tubules. It is the silken cocoon of butterflies and moths that is commonly called a *chrysalis*, in admiration of the glistening golden appearance of the more spectacular kinds.

Among flies, those more advanced groups that have evolved the maggot as a larva – among them the house fly and the bluebottle – have achieved the protection of a cocoon by not breaking out of the last larval skin. This detached skin, hardened and darkened, forms a seed-like object known as a *puparium*. Textbooks often refer to these as a third type of pupa, the *coarctate* pupa, but this is incorrect. The puparium is a type of cocoon, not a type of pupa. Inside it the real pupa is always exarate.

The formation of a cocoon is an adaptive modification, that is a form of protection against the dangers of the world outside. Many different insects have evolved a cocoon, independently of each other. For example the 'ant-eggs' that are used as food for fish in small aquariums are not eggs, but pupae in a cocoon. On the other hand the pupae themselves are believed to have many hereditary features which may throw light on the ancestry of the various groups of holometabolous insects. Hinton [42] considered that a basic division exists between pupae with movable mandibles (decticous pupae) and those in which the mandibles no longer move (adecticous pupae). His paper, *A New Classification of Insect Pupae*, besides expounding this view, gives a great deal of information about the pupae of various groups of insects.

3
The primitive insects

The major groups into which insects are divided for discussion and comparison are the *orders*, a list of which is given in Table 3 (Appendix, p. 281). It should be remembered that an order of insects is the zoological equivalent of an order of mammals, say the rodents or the carnivores. Some orders of insects are readily recognised by non-entomologists: for example all the many kinds of beetles belong to the order Coleoptera, and all the butterflies and moths to the order Lepidoptera; all Dragon-flies are Odonata, and all fleas Siphonaptera. Some orders, such as Hemiptera ('bugs') and Diptera ('flies') use a commonplace name in a precise sense, and so exclude many insects that a layman would refer to as 'bugs' and 'flies' respectively. Yet again, there are many obscure insects that do not seem to have any close relatives, and so need to be placed in separate small orders. To accommodate all these it is necessary to arrange the insects in about thirty orders.

Even then there is not complete agreement among entomologists, and textbooks seldom give exactly the same list. That given in Table 3 is very close to the list given in Imms's *General Textbook of Entomology*, a standard work of reference. A number of orders are known only from fossils, having become extinct during the long period since the fossils were preserved. In an attempt to give as coherent a picture as possible of the evolution of insects, these extinct orders are mentioned below in their logical sequence, and their names are marked with a dagger (†). It will be noticed that all these extinct orders come fairly early in the evolutionary sequence of insects. No order of insects has become extinct for a very long time.

Sub-class Apterygota

These insects are believed to have been wingless from their earliest ancestry, in contrast to those other wingless insects that are believed to have lost their wings in evolution. All Apterygota are

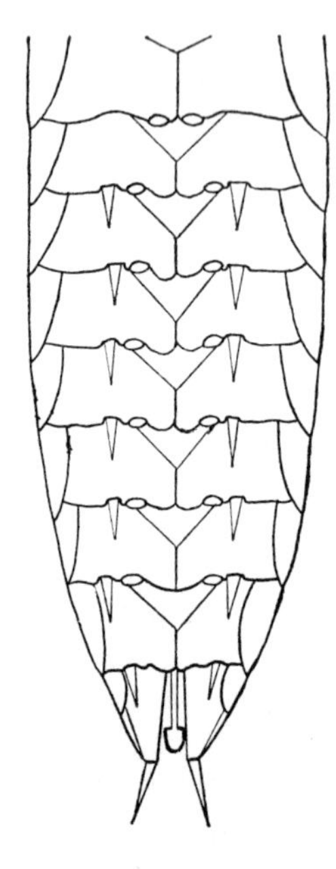

Figure 13. Abdominal styli of an apterygote insect (after Delany) (see also figure 6 and plate 2).

small, and some are minute. They are usually divided into four orders as follows:

1. *Order Thysanura* (350 *species*)*

Collectively known as bristletails, of which the most familiar are *Lepisma*, the silverfish, and *Thermobia*, the firebrat (plate 2), which appear in hearths and kitchens at night, and scurry away with a twisting motion when the light is switched on. *Machilis* is a

* The number given in brackets after the name of each order is a rough estimate of the number of species at present known to occur throughout the world. This has little or no value in itself, because anyone who studies any of the larger groups of insects can quickly discover new species. Nevertheless these figures give some indication of the comparative size of the order. Often an order which now contains only a few species has passed its evolutionary peak and is declining, but this is not always true. Strepsiptera, for instance, are peculiar parasitic insects, and it is unlikely that they were ever very abundant.

large insect of similar shape, found outdoors under stones and dead wood. *Petrobius* also occurs among rocks on the seashore. Thysanura have biting mouthparts which project from the head (ectognathous).

This order of insects is considered to be the most primitive and nearest to the ancestral type. In a number of ways Thysanura show how the structure of more advanced insects may have arisen. The thoracic legs are simple, with only a few joints, and in *Machilis* and *Petrobius* there are *styli*, vestiges of a primitive 'biramous appendage' common to other classes of arthropods, such as Crustacea.

2. *Order Diplura* (400 *species*)

These are superficially rather similar to the Thysanura, and were

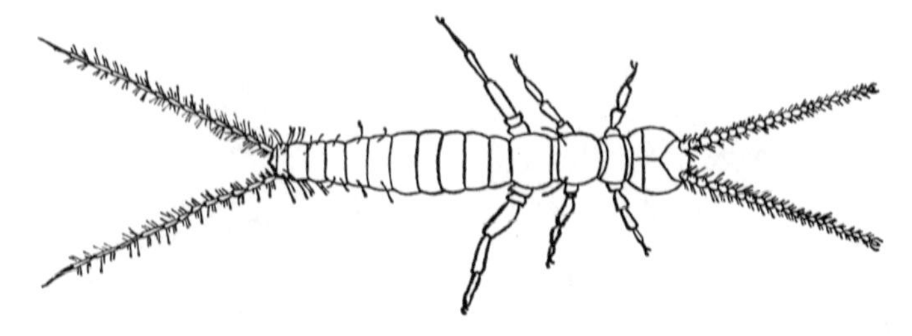

Figure 14. Adult of the primitively wingless insect *Campodea* (order Diplura). Larvae of winged insects that run actively on three thoracic legs in a similar way are called **campodeiform larvae** (after Delany).

formerly placed in the same order. In some ways they are more advanced, in particular in having the mouthparts withdrawn into the head (entognathous). They have abdominal styles, or rudimentary legs, but no eyes. One of the genera in this order is *Campodea*, after which the Campodeiform larva is named; others are *Japyx*, *Projapyx* and *Anajapyx*.

3. *Order Protura* (100 *species*)

Tiny insects, little more than one millimetre long, which were not discovered until 1907 and which even in 1955 were sufficiently rare for living specimens to be specially exhibited and discussed at a meeting of the Royal Entomological Society of London. They are sometimes called 'telsontails', in reference to an alleged post-segmental region called the *telson*, which is present in Arachnida,

Crustacea and Myriapoda, but only as an embryonic rudiment in insects; Protura have something resembling this at the tip of the abdomen. Protura have no common name.

Protura also have the mouthparts withdrawn into the head, and other specialised features, such as the loss of the eyes and antennae, and sometimes of the entire tracheal system. In these matters they must have turned aside from the main line of insect evolution at an early stage. They make up for the loss of antennae by using the first pair of legs as sensory organs, in a way comparable to that of some Arachnida. They are thus a specialised sideline, and scarcely insects at all. A further divergence from true insects is that they start life with nine abdominal segments, and add one more at each of their three moults, to reach a total of twelve.

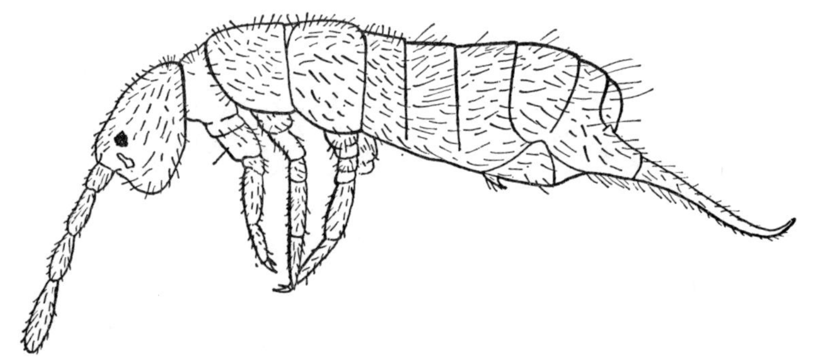

Figure 15. A spring-tail, a primitively wingless insect of the order Collembola.

4. *Order Collembola* (2,000 *species*)

Spring-tails, so-called because they possess a *furcula*, a forked organ which turns forwards underneath the abdomen and is gripped by another structure called a *tenaculum* (figure 15). When this is released the insect is projected upwards and forwards, like those model frogs made for children.

Though Collembola are tiny, mostly less than five millimetres long, they are varied and extremely abundant. Soil, leaf-mould and other detritus contains uncounted millions of individuals, which can be extracted by means of the Berlese funnel. Others live on the sea-shore and are repeatedly submerged by the tide. Even more than Diplura and Protura, Collembola have diverged from the main line of evolution. They also are entognathous, with mouthparts withdrawn into the head. They have antennae but no eyes, and are usually without either tracheae or Malpighian tubules (intestinal excretory organs). The number of abdominal segments is reduced to six. The mouthparts of some Collembola emphasise the mandibles and are clearly used for chewing, while others have sucking mouthparts. They always live in moist places, and probably feed on fungoid growths. South [99a] found that the Collembola (*Entomobrya*) could be reared entirely on the fungus *Cladosporium*.

Some authorities consider that Collembola diverged before true insects had been evolved, but this obviously depends upon where we decide to draw a line and how we define insects. If they are not insects then we have to find some other group to put them in, and little is gained by doing that.

Sub-class Pterygota

There are about three thousand species of primitively wingless insects in the group that has been discussed. All other insects, whether they have wings or not, belong in the sub-class Pterygota. Since almost three quarters of a million insects are known, we can say that pterygote insects are overwhelmingly predominant, in the ratio of more than 250:1.

The view that the Apterygota came first in evolution receives rather dubious support from the fossil record. The most ancient remains that can be identified as insects are from a fossil peat-bog at Rhynie in Scotland, and date back to Middle Devonian times. Here there are traces of what appear to be Collembola, and, moreover, of Collembola very much as we know them today. This is a recurrent problem in considering fossil animals, that the earliest known form is still very advanced. It must be presumed that much, perhaps most, of its evolutionary history was already behind it, even at that remote date.

Winged insects make their first appearance in deposits of the Carboniferous period, among the steamy swamps of the Coal Measures. Perhaps these conditions, under which it must have been difficult to move about on the ground, set a premium on flight. It is generally considered that wings arose in insects first as horizontal pads, a pair on each of the three thoracic segments. Perhaps these acted as gliding planes, like the membranous skin of flying squirrels of today. Tropical swamps, with tall, leafy trees, favour the evolution of insects which can fall relatively slowly, and can prolong their glide so that their forward speed carries them to the safety of another tree. A slender, elongate insect with a big surface area for its weight, increased first by thoracic pads, which first increase enormously in area and then become capable of movement – these are trends that seem to suit such a setting, and which in fact are those of the earliest known insects (see also Chapter 10).

It is characteristic of these earliest known insects that when at rest on vegetation or on the ground they hold the wings stretched

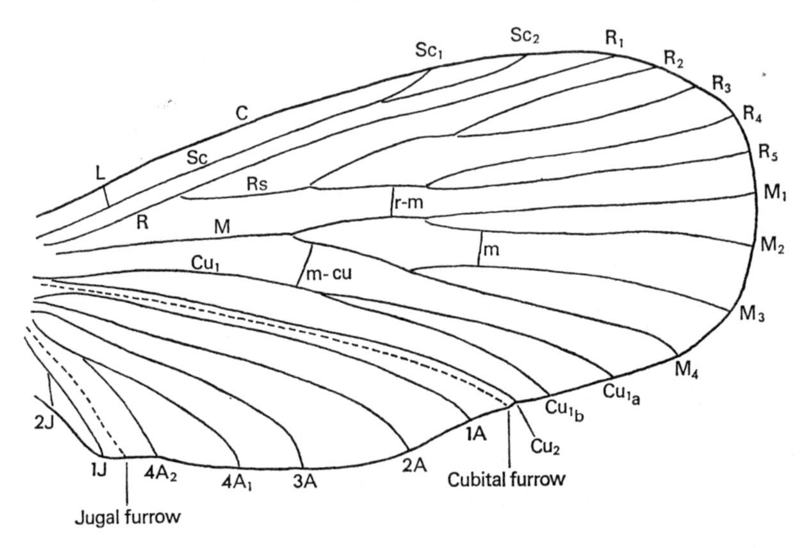

Figure 16. A primitive arrangement of the wing-veins, named after the Comstock-Needham system (after Ross).

out flat, at right angles to the body, and have no means of folding them. This is typical of the Dragon-flies of today, and a reason why their wings catch the light and show off their iridescent sparkle so well. May-flies do the same, but all other living winged insects fold their wings in various ways. Insects of this early group are called Palaeoptera, i.e. with the old type of wings.

Section Palaeoptera

The long Carboniferous period was the heyday of such insects, and wings of some magnificent specimens have been preserved as fossils. Some had a wing-span of upwards of twenty inches (fifty centimetres). The order Palaeodictyoptera† had chewing or sucking mouthparts, and are believed to have fed on vegetation, either growing or decaying, of which there was an abundance. In addition to two pairs of wings, some of the earlier members of this order had wing-pads, or paranotal lobes, on the prothorax: opinion differs as to whether or not these could be moved. Even the segments of the abdomen had their pairs of lobes. This is evidence from fossils in favour of the theory that in origin the various segments of an insect's body were all similar, and became different in the course of evolution.

The deposits of the Upper Carboniferous period show that at that time there were also present two other orders, Meganisoptera† and Protephemeroptera† which had already begun to advance towards the insects we know today. In particular, they had lost the paranotal lobes and had simplified the venation of the wing to some extent further still. They were still big, and *Meganeura monyi* of the Upper Carboniferous was the biggest with a wing-span of nearly thirty inches (seventy centimetres).

The hot, humid Carboniferous period came to an end at last, and the succeeding Permian period was one of increasing drought. This was the time when vertebrates took to the land, the amphibia of the swamps being largely replaced by the reptiles, with their greater efficiency as land-animals. A comparable development occurred in insects with a great stimulus to the evolution of more versatile types, less completely dependent on water and swamp vegetation. Some water remained, of course, in which lived on not only the surviving amphibia but also, among insects, the survivors of the Protephemeropters and Meganisoptera, from which evolved two orders of insects that still exist today: the Dragon-flies and May-flies.

5. *Order Odonata* (5,000 *species*)

Dragon-flies are familiar to everyone, in all parts of the world. The name 'Dragon-fly' is appropriate to all of them, since they live by attacking and eating other insects, but strictly speaking it is applied only to one of the three sub-orders, Anisoptera. These are the ones with 'unequal wings', that is with the hind-wings broader than the fore-wings (plate 20), generally the biggest and most aggressive insects of this order. The sub-order Zygoptera, with 'paired wings' are called 'Damsel-flies', and are more delicate, slender insects, generally smaller. Damsel-flies hold the wings at rest with the two wings of each pair back to back, whereas true Dragon-flies hold them out at right angles to the body. Two rare, isolated species form a third sub-order Anisozygoptera.

Odonata, like their ancestors, still have aquatic nymphs, which are also predatory, and which have evolved an efficient food-capturing organ called the 'mask': the labium is hinged, and capable of being shown forwards for as much as one third of the body-length of the nymph (figure 62). As more direct adaptations to aquatic life the nymphs have gills, which are discussed more fully, with other such adaptations, in Chapter 12. The nymphs take every kind of aquatic animal life, often minute, but the larger nymphs will take worms, tadpoles and even small fish.

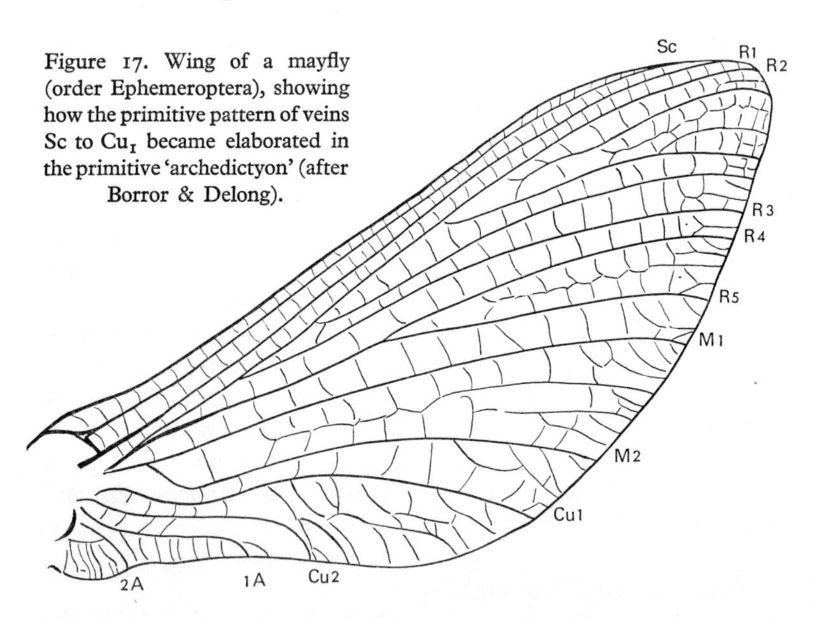

Figure 17. Wing of a mayfly (order Ephemeroptera), showing how the primitive pattern of veins Sc to Cu_1 became elaborated in the primitive 'archedictyon' (after Borror & Delong).

6. *Order Ephemeroptera* (1,500 *species*)

May-flies are also still fully tied to the waterside with an aquatic nymph that breathes by gills. Whereas in Dragon-flies the nymphs and the adults are equally voracious, May-flies have divided the basic function between immature and adult stages. All feeding is carried out by the nymphs, which feed mainly on underwater vegetation and may live for a year or more. Adult life is very short, a matter often of only a few hours, or at most a few days. Hence the scientific name of the order, which means 'winged insects that live for a day'. The business of the adults is to mate and lay eggs.

The adult May-fly is a less powerful flier than the Dragon-flies, but takes part in impressive mating flights or swarms, over or near water. The hind-wings are much reduced, and illustrate the tendency common among winged insects, to rely mainly upon one of the two pairs of wings for flight, or else to link the two wings of each side so that they function as one. May-flies show a curious mixture of archaic features and advanced evolutionary ideas. They are unique in expanding the wings before the final moult, and so having a *subimago* before the *imago*, or true adult. Fishermen recognise the subimago as a 'dun', and the imago, glistening after shedding its membranous covering, as a 'spinner'.

4
Orthopteroid insects

All other insects are grouped in the section Neoptera, or those with wings of the 'new type', because they fold their wings back when they come to rest, so that the fore-wings cover the hind-wings. Among Palaeoptera, the Damsel-flies of the order Odonata hold their wings up above the body, back to back, but they are unable to turn so that they lie flat over the abdomen. In this rotation of the wing the insects had made a significant step forward in evolution, not only making the wing less cumbersome when it was not in use, but paving the way to more subtle movements of the wing during flight.

A further characteristic of Neoptera is that the wings that do most of the work in flight – usually the hind pair, but fore-wings are important in Hemiptera and paramount in Diptera – have an extended area of membrane at the posterior angle of the base (figure 18). This is called the *jugal* fold, or the *neala*. The locust and the cockroach both show this 'new wing' highly developed: if the hind-wing is drawn out to its full extent, a great fan-like area, much pleated, is revealed. If the principal veins of the hind-wing are labelled on the Comstock-Needham system (see figures 16 & 18) it will be seen that this jugal field lies beyond the anal area of the primitive wing, and is thus a new structure, with its own system of radiating veins.

Neoptera can be classified into twenty-four orders. (Table 3), which can be further divided into Exopterygota (Hemimetabola) and Endopterygota (Holometabola). The difference in metamorphosis between these two has been explained in Chapter 2.

Hemimetabola (Exopterygota)

The two orders of Palaeoptera discussed in Chapter 3 – Ephemeroptera and Odonata – are Hemimetabola, along with a further fourteen orders of Neoptera. These fourteen may be considered as grouped into two super-orders: Orthopteroidea, relatives of the Orthoptera, or grasshoppers; and Hemipteroidea, relatives of the Hemiptera, or bugs.

Super-order Orthopteroidea are essentially primitive insects, with mouthparts of a pattern suited for chewing and masticating food. They have 'old fashioned' wings, with many veins, and often with many crossveins, and other veins supernumerary to the basic system (figure 16). Thirty years ago most of them would have been included in a single order Orthoptera, but modern classifications divide them among no fewer than nine orders, as follows:

7. *Order Plecoptera* (1,500 *species*)

Stone-flies: Perlaria. These are aquatic insects rather like May-flies, but with the more advanced type of wing characteristic of Neoptera. Like Ephemeroptera and Odonata, Plecoptera have aquatic nymphs, which have a closed tracheal system, and breathe through tracheal gills. Another archaic feature of Stone-flies is their

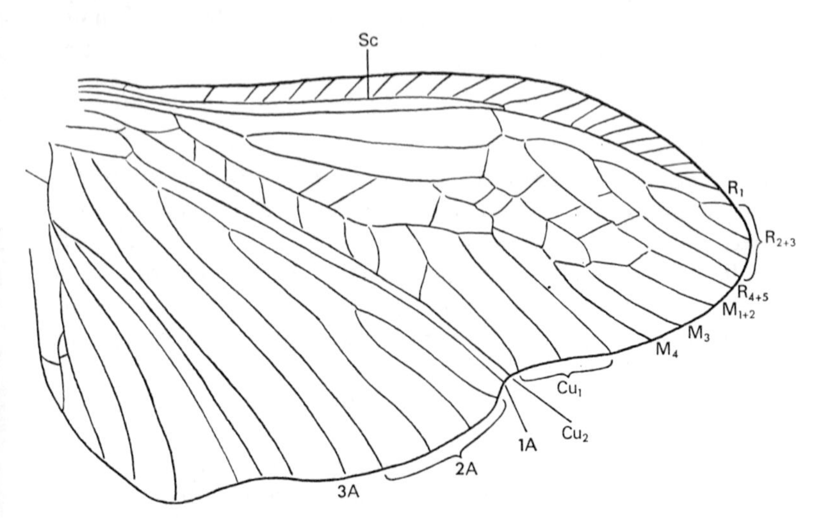

Figure 18. Hind-wing of a stone-fly (order Plecoptera) showing how the primitive system of venation may be modified.

long nymphal life, of one to four years, coupled with a large number of moults, up to thirty-three having been recorded.

The name 'Stone-flies' is given to them because they are usually found on stones near water. They are soft-bodied, weakly flying, furtive insects, which fly little by day, though sometimes they come in large numbers to a light at night. Both nymphs and adults have chewing mouthparts: nymphs may be either vegetarian or carnivorous; adults feed on blue-green algae, and some do not feed at all in the adult stage.

8. *Order Grylloblattodea* (*Notoptera*) (5 *species only*)

These few rare animals are found in mountainous areas of North America, Japan and Russia, where they live at the edge of glaciers. They are wingless, elongate insects, fifteen–thirty millimetres long, with long antennae composed of up to forty segments. Structurally they combine features that are otherwise found separately in different orders of the Orthopteroidea. For instance they have a pair of cerci (figure 14) like those of cockroaches, but combine these with a sword-shaped ovipositor like that of the long-horned grasshoppers.

Both the structure and habits of Grylloblattodea support the view that these are another group of archaic insects, and indeed one on the verge of extinction. It would seem that only in such inhospitable places can they survive the competition of forms more advanced in evolution.

9. *Order Orthoptera* (10,000 *species*)

In the restricted sense in which this name is now used it includes only the grasshoppers and crickets, other orders having been created for the cockroaches and other insects that were formerly included. There are three superfamilies: Acridioidea, the locusts and short-horned grasshoppers (i.e. those with short antennae and little or no ovipositor); Tettigonioidea, or long-horned grasshoppers (with long antennae and long, sword-shaped ovipositor); and Grylloidea, or crickets (with long antennae and slender, cylindrical ovipositor).

All three groups are jumping insects, with the hind legs longer than the others. The head is held downwards, with the mouth pointing ventrally (plates 13–18), and the mouthparts are of a generalised, chewing type. As a general rule these insects are terrestrial and plant-feeding, but some have gone back to the water, and some have become carnivorous. For example, some of the Tetrigidae, or Grouse Locusts, have the hind legs adapted for swimming and there is a water-skating cricket: while certain members of the family Gryllacrididae feed on aphids, a logical development from plant-feeding that has occurred a number of times in the evolution of insects.

Another characteristic of this order is the production of sound by *stridulation*, or rubbing one surface against another. Locusts and short-horned grasshoppers usually rub a roughened femur of the hind-leg against a hardened area of the wing; whereas long-horned grasshoppers and crickets rub the two wings together (see Chapter 9 for a fuller account).

Acridioidea are the familiar grasshoppers of grassland and scrub. Locusts are certain grasshoppers that have developed the trick of mutually exciting each other until they move off in a mass flight, devastating the vegetation of any area upon which they alight to feed. The habits of locusts are discussed in Chapter 20.

Acridioidea are 'grazing' insects, and especially feeders on grass, as the name grasshopper implies. Tettigonioidea are called Katydids in North America in reference to the sound that they seem to make when they are stridulating. They are tree-living insects, and therefore analogous to 'browsers' rather than 'grazers'. Gryllidae, the crickets, are structurally related more to long-horned than to short-horned grasshoppers, and are celebrated for their penetrating chirp or 'song', which they produce by rubbing the wings together. In habit they are more like cockroaches than grasshoppers, and live among any kind of debris, indoors or out. The mole-crickets (*Gryllotalpa*) burrow into the ground with the aid of very highly developed fossorial fore-legs (plate 14).

10. *Order Phasmida* (*Cheleutoptera*) (700 *species*)

The stick-insects, walking sticks and leaf-insects are now placed in an order of their own. These are Orthopteroid insects that are most highly modified in a way that gives them the maximum of concealments and protection against enemies. The stick-insects are elongate, and often very long, a few of them reaching thirteen inches or more. The legs are uniform, i.e. suited not to jumping but to running or, rather, to walking. The meso- and metathorax are elongate, the prothorax small. Though many stick-insects can fly, only the hind pair of wings is ever well-developed, and even these are absent in many species. Both in colour and texture the surface of the body strongly resembles a living or dead twig, and as long as the animal remains motionless it is almost impossible to detect.

These are sluggish insects, moving little and feeding on the vegetation among which they live. Perhaps because they are so sluggish, and make so few demands on their environment, stick insects are easy to rear in the laboratory. Their natural habitat is throughout the warmer countries of the world.

Leaf-insects (Phyllidae) are members of this order that have specialised in a slightly different direction. They still rely for protection on concealment, but resemble leaves rather than twigs. The females, in particular, have broad expanded areas of the legs and abdomen which, together with the expanded fore-legs, are coloured and marked in a way that resembles a collection of leaves. They are mostly found in the humid tropics of the Oriental region.

11. *Order Dictyoptera* (1,200 *species*)

The cockroaches are familiar to every student of entomology as the most usual 'type' upon which to study the structure of insects. In many ways the cockroaches deserve this distinction, since they are an ancient group, known as fossils from the very beginning of the Carboniferous period and, what is more, they were then very much as they are today. This means two things; that they must have carried out the progressive part of their evolution long before then, and that they must be very efficiently adapted to their way of life if they have been able to survive unchanged for the last four hundred million years.

The most characteristic feature of cockroaches is the enlargement of the prothorax into a shield which covers the head. The fore-wings are leathery, and serve as covers for the hind wings (plate 13), while the latter have a great development of the fan-like neala, or anal fan. These are developments such as can also be seen in the earwigs, but in the cockroaches they have not progressed so far. Female cockroaches often have reduced wings, or none at all. The cerci are well-developed, but not to the extent of the forceps of the earwigs.

Cockroaches are general feeders. They have chewing mouthparts, often illustrated in textbooks as the primitive example of

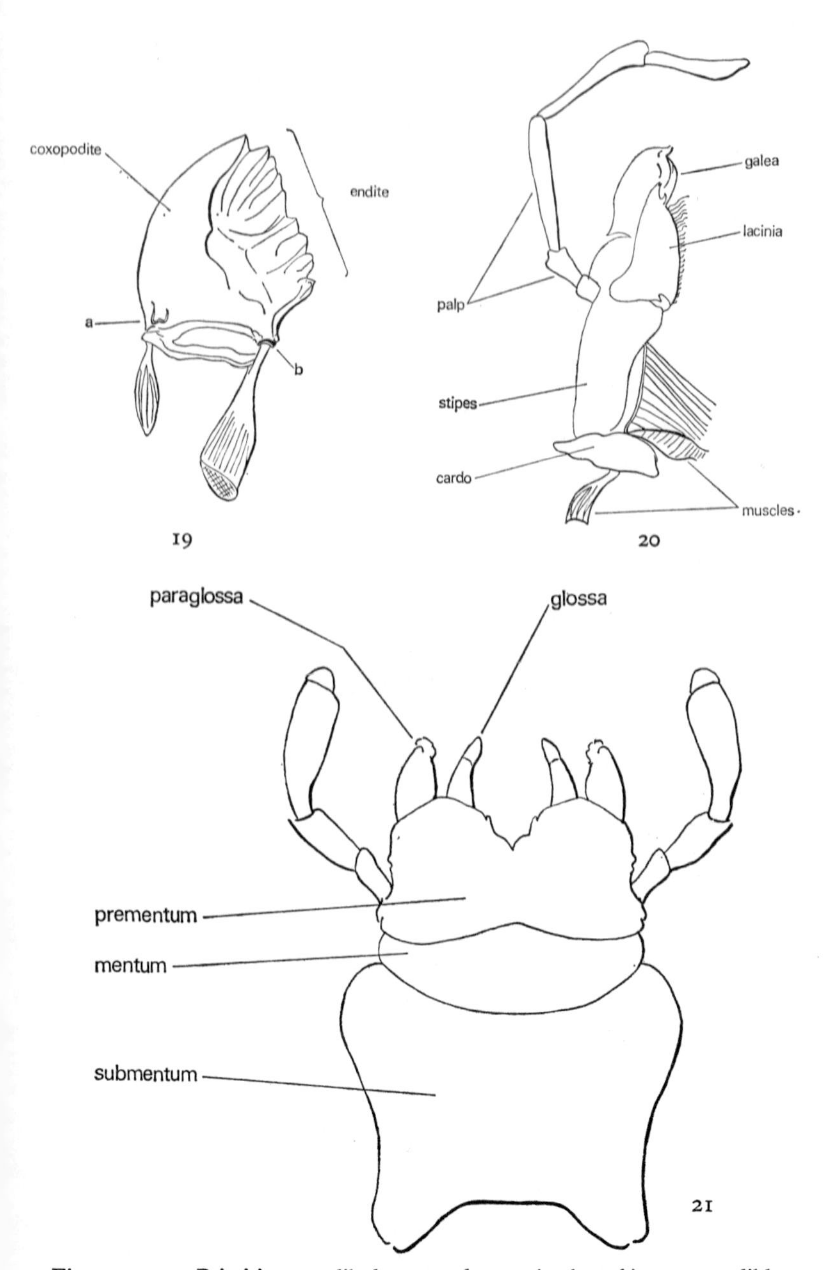

Figures 19–21. Primitive mandibulate mouthparts (cockroach): 19, mandible; 20, maxilla; 21, labium, a single structure formed from a pair of appendages. All other types of mouthparts found among insects can be interpreted as variations from these by elaboration of some parts and suppression of others. a, b, pivots of mandible.

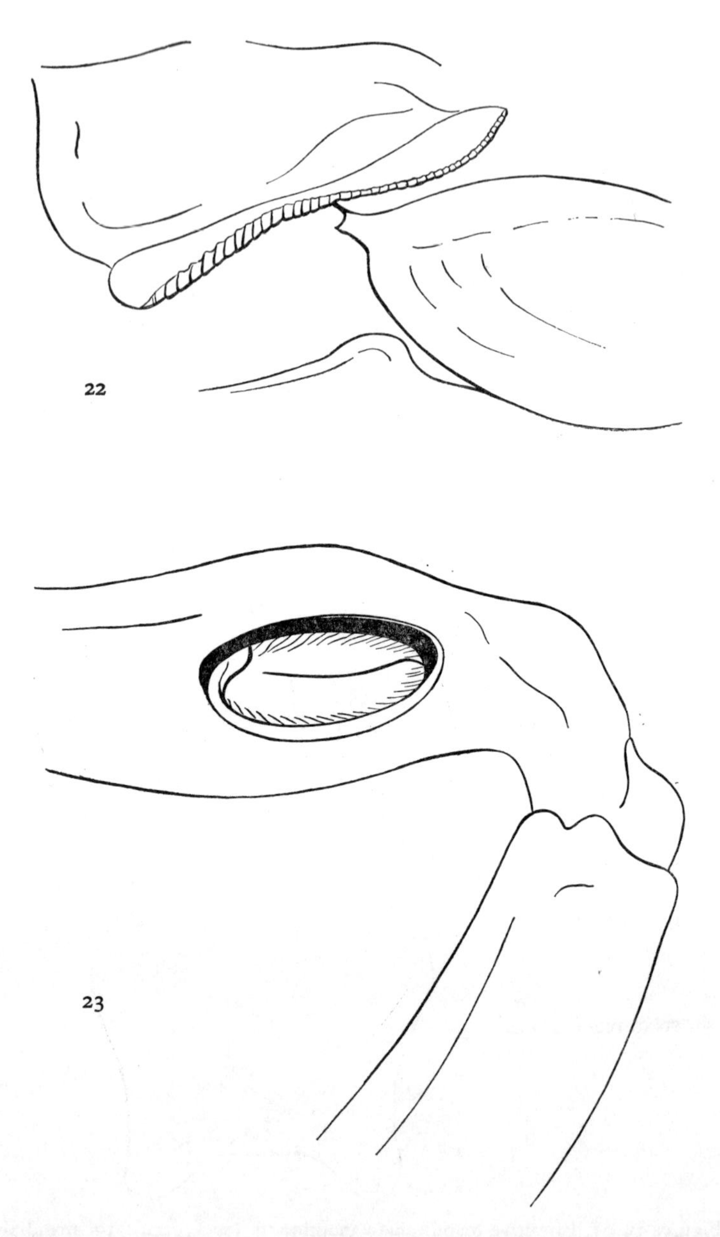

Figures 22, 23. Two of the very large number of organs concerned with the production and reception of sound: 22, file and scraper of a katydid (order Orthoptera); 23, hearing organ on the leg of another katydid (after Pierce).

their kind. One thinks of cockroaches as household pests, though in fact only one or two species regularly live in man-made environments. Nearly all cockroaches live out of doors, under leaves and other debris, or on the vegetation. They are predominantly tropical insects, and the house-infesting species live in the artificial tropics of kitchens, bakeries and other heated buildings. They are sometimes called 'steam-flies' presumably because they appear from behind the steam-pipes of heating systems, and 'black-beetles' although they have no close relationship with the true beetles.

They do not carry disease, as far as is known, but they crawl over food, fall into soup, and so on, and are generally unwelcome. The three common household cockroaches are *Blatella germanica*, the German Cockroach, or Crotonbug; *Blatta orientalis*, the Oriental Cockroach; and *Periplaneta americana*, the American Cockroach. The last is a big, reddish-brown insect, about two inches long.

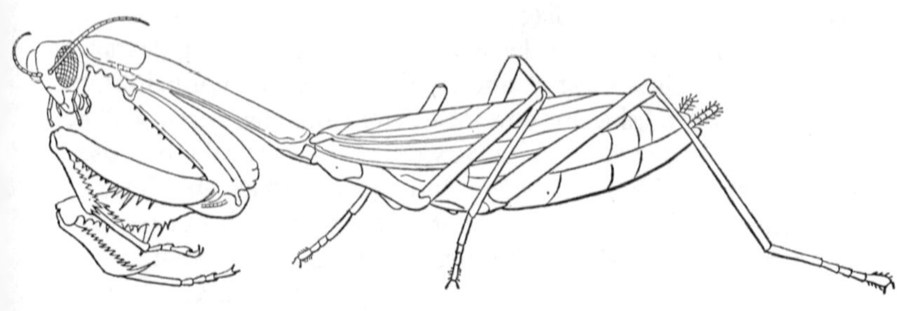

Figure 24. A praying mantis (order Dictyoptera) (after Ross).

Mantids are rather curiously linked with cockroaches, even in the stricter classifications of the present day. Superficially they are very different in appearance (figure 24), the whole front end of the insect being highly modified for seizing and holding its prey. The prothorax is greatly lengthened, and raises up the head and fore-legs. The fore-legs have strong spines on the lower side of the femora, and the tibiae can be closed against these. The mouthparts are mandibulate and very efficient, and the head, often with huge eyes, can be turned in almost any direction on the highly mobile neck.

They are called 'praying' or 'preying' mantids because when they are waiting motionless for possible prey to come within reach they seem to adopt an attitude of prayer. They are very voracious, and cannibalistic, and the bigger females will even devour the male during the act of mating. They have much of the concealing coloration of stick insects and leaf-insects, and are difficult to see in their natural setting, until they move.

Mantids are linked with cockroaches in one order partly because they lay their eggs in batches enclosed in an egg-pod, or *ootheca*.

12. *Order Dermaptera* (1,200 *species*)

Earwigs are common insects of the garden, and come into houses fairly often. They are chiefly nocturnal, and will come to a light at night; during the day they hide away in crevices of all kinds, and are easy to decoy into the familiar 'earwig traps' made from inverted flower-pots and such-like objects. In spite of their common name they do not invade the human ear, except by chance. They are said to be omnivorous, eating anything that they can chew, and examples are known of their eating small insects, and of their damaging flowers and foliage.

These leathery insects have a curiously archaic appearance though in fact their most characteristic features are not primitive, They have large 'pincers' developed from the anal cerci, and no one knows exactly what these are used for. They are said to be used at times when folding the wings, and again in attack and defence. The latter use is probably more apparent than real, but perhaps the forceps have a value in frightening off potential enemies.

The hind-wings of earwigs are certainly not primitive. They are

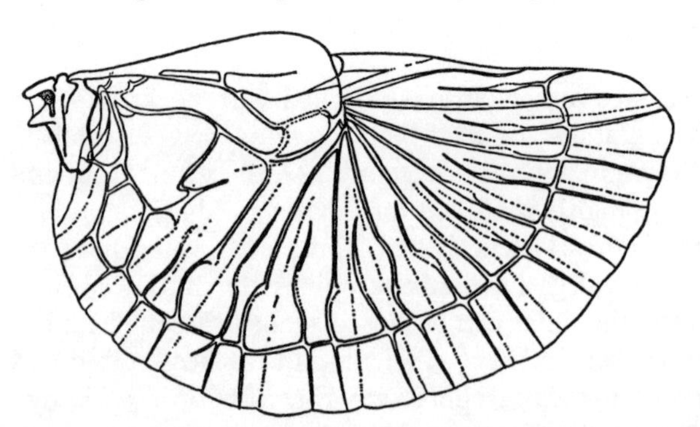

Figure 25. Hind-wing of an earwig (order Dermaptera), highly adapted mechanically to fold up like a fan, with consequent loss of all resemblance to the primitive pattern of Figure 16 (after Ross).

much pleated in a radial, fan-like manner (figure 25), and this part is the neala, or new wing behind the anal area of the primitive wing (cf. page 39). The 'old wing' is reduced to a vestige. This development indicates that earwigs have evolved along a line of their own, away from the main stem of the Orthopteroid orders. Their name of Dermaptera applies to the fore-wings, which have lost all venation and become wing-cases, just as they have in beetles. Indeed, in this respect earwigs have evolved along lines somewhat parallel to those of beetles.

Although Dermaptera have remained primitive in their metamorphosis, with little change during the life of the insect, they are an ancient group going back to the Jurassic, with ancestral Protelytroptera† in the Permian. They have had time to evolve a number of wingless and semi-apterous forms, including two genera that have exceptional life-histories. *Arixenia* is a genus of viviparous insects, wingless and almost without eyes, which live with and upon the Indian bat *Cheiromeles torquatus*. *Hemimerus*, also viviparous, wingless and blind, lives in the fur of the Giant Rat, *Cricetomys gambianus*, in West Africa. These modifications, which suit a parasitic existence, or life in a dark cave, are common throughout the insects, and can be seen among even such advanced insects as the true flies, or Diptera.

Another advanced feature of earwigs is that they take care of their eggs, and even of the young nymphs. Maternal care is rare among insects, apart from the fully social ants, bees, wasps and termites – and even there it is hardly 'maternal' care, but rather the provision of insect nursemaids (see Chapter 19).

13. *Order Isoptera* (1,700 *species*)

Termites, with their highly elaborate social life, are a unique development so far back in the scale of insect evolution, though rudiments of maternal care exist in Dermaptera. A degree of social organisation comparable to that of termites is not to be seen again until the Hymenoptera, the bees, wasps and ants, which appear to be a very much later evolutionary experiment.

Adult termites are pale, soft-bodied creatures, with relatively long wings that are held horizontally over the body when at rest. The wings of the two pairs are similar in size and venation, and the veins are a simple series of forks. Some primitive termites have an archedictyon, the primitive network seen in May-flies and others, but the normal cross-veins are not present. Presumably the wings have been simplified in evolution, perhaps in correlation with the fact that reproductive male and female termites use their wings only for a brief mating flight, and afterwards shed them by breaking them off at the base; workers and soldiers are wingless.

Although termites are grouped among Neoptera they are almost without the neala or jugal field of the wing. This lobe is well-developed in the Australian *Mastotermes*, considered to be a primitive genus, and this fact gives support to the idea that in other termites the neala has been lost.

Termites are often called 'white ants': they can be distinguished from true ants because the latter, besides being more heavily sclerotised, and harder, have the hind-wings distinctly smaller than the fore-wings, and both wings have the characteristic cells of the Hymenoptera (cf. figure 42). Worker ants, like worker termites are wingless, but can be recognised by having the antennae bent into an elbow, and the abdomen strongly 'waisted', with a knob-like structure at the narrowest part.

To return to termites, these are social insects and always live in communities. Chapter 15 discusses their social life in more detail and compares it with the social life of Hymenoptera. For the present it is sufficient to note that termites are *polymorphic:* that is that several different kinds of individual exist within one species, and indeed within one community, and – most important – each kind, or *caste*, takes a different share in the duties of the community. The sexually mature males and females are called *reproductives* ('kings' and 'queens'). They have functional wings at their final moult, but shed these after making a mating flight. The males remain small, but the fertilised female nest-builder, or queen termite, once established in her nest, becomes distended with a mass of eggs until her abdomen is many times bigger than the thorax, and is a mere membranous bag. She is fed and tended by wingless, sterile individuals of either sex. Other sterile individuals have a defensive rôle as *soldiers*, and have the head enlarged and hardened, with either large mandibles (mandibulate soldiers), or with a long, nose-like rostrum (nasute soldiers). There are also supplementary *reproductives*, males and females able to reproduce, and capable of either supplementing the efforts of the king and queen, or of replacing them altogether. They have short wings or none at all (plate 41).

The nests of termites are varied and often very elaborate, but fall broadly into two kinds. As their name suggests, 'dry-wood termites' live in living or dead wood above ground level, and are highly destructive to wooden buildings. In the warmer countries in which termites mostly occur their activities often cause wooden buildings to collapse suddenly. The second group, the subterranean termites, form colonies underground, but often extend their nests above the ground, and thus create the well-known *termitaria*. Strictly speaking, any place inhabited by termites is a termitarium, but it is convenient to use this name for the structures visible above the ground because it is less inaccurate than the common name of 'ant-hills'. These may vary in height from a few inches to twenty feet. The biggest of them are as hard as concrete, and become very durable features of the landscape.

Biologically, the termites are of special interest for their social habits; for their polymorphism, or existence in several castes; and for their ability to digest cellulose, and therefore to feed on the actual tissue of wood. Other wood-boring insects, such as beetles, do not digest the cellular tissue of the wood, but only the sugars and other carbohydrates that it contains. Even termites do not digest wood without assistance. In the hind gut they have a colony of single-celled (or acellular), flagellate Protozoa, which break down cellulose into carbohydrates that can be absorbed and digested by the termite. These Protozoa are present in workers and soldiers, which eat wood, but are not present in king or queen termites, which are given a special diet by their attendant workers. Nor are they found in some of the most advanced termites, which no longer feed on wood, but on fungi.

14. *Order Embioptera* (150 *species*)

These are an evolutionary sideline of small, fragile, elongate insects, which live furtively under stones, under bark, or among mosses and other concealing vegetation. They have a curious mixture of primitive and specialised features. They have a cylindrical body, elongate antennae, and well-developed cerci, and the wings, which are present only in males, are almost alike in the two pairs. The veins are simplified down to the major branches, sometimes reinforced with a few cross-veins. The females have no wings, and do not change in external form from the time of hatching from the egg: that is, they have no apparent metamorphosis.

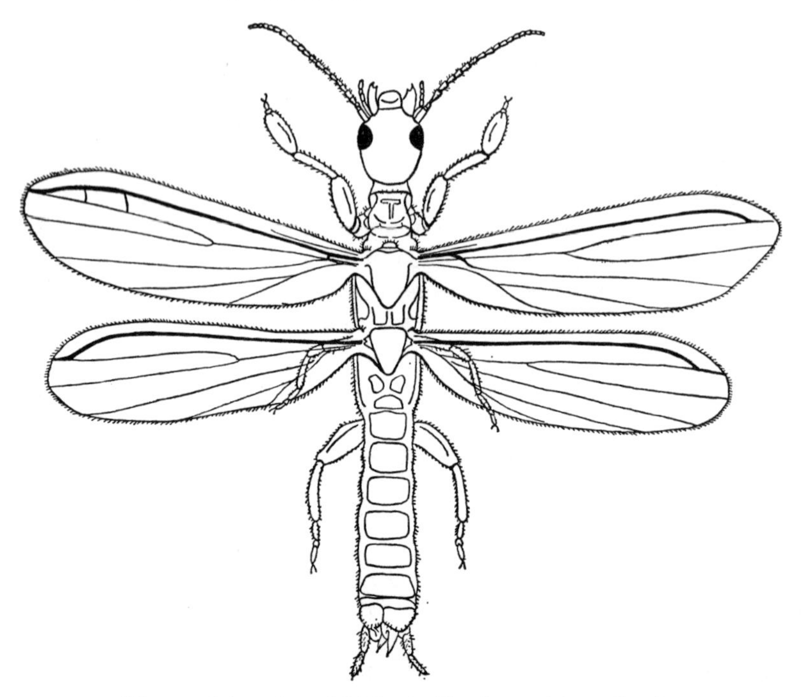

Figure 26. A male embiid (order Embioptera) (after Ross).

Embioptera are primitive in the elongate body, the three almost equal thoracic segments (figure 26), and some of the internal structure, especially that of the reproductive organs, which still show traces of a primitive segmental arrangement. On the other hand, the pale, poorly sclerotised body, the small eyes, the reduced rather ineffective wings of the males, and the winglessness of the females – all these are obvious adaptations to the furtive life they lead. They spend their lives in tubes made from silken threads, which they spin from glands in the first segment of the fore tarsus. Textbooks sometimes refer to Embioptera as 'web-spinners', though they are really not well enough known to have a common name.

Embioptera are found in all warm countries, tropical and subtropical, including Australia, and on many islands, though they are always few and rare. Such a distribution, spread thinly over the whole area where the climate is suitable, suggests a group of some antiquity, that has adapted itself efficiently to a small biological niche. There will always be such places where Embioptera can remain concealed, and it seems likely that, although they are few in number, they are not in danger of extinction.

15. *Order Zoraptera (about 20 species only)*

Minute insects, with no common name. They are pale, rather like small termites, and they live mainly in colonies, though not sharing an organised social life. They are interesting to the student of evolution because they provide a link between the Orthopteroidea that we have just examined and the Hemipteroidea that we are about to consider. The head, with its long antennae, the cerci, and the general body shape, suggest that these are Orthopteroid insects, almost as much so as the termites. Yet in internal structure they have a concentration away from the primitive segmental plan, and are apparently close to the order Plecoptera with which the next super-order begins. Indeed Zoraptera were formerly considered to be an obscure branch of Psocoptera, but recent opinion regards the two as separate orders arising from diverging branches of the evolutionary stem, one of which led to the Orthopteroid orders, and the other to the Hemipteroidea.

Zoraptera are placed in a single genus, *Zorotypus*, and the few species, like those of Embioptera, are scattered all over the warmer countries of the world.

5
The Hemipteroid orders

The remaining five orders of Hemimetabolous insects are considered to derive from a separate evolutionary stem. Starting off in the primitive way as insects that bite and chew their food, three of these five orders have modified the mouthparts into a tubular proboscis which is adapted for piercing either animal or plant tissues, and sucking up liquid food. We begin with what is apparently the most primitive.

16. *Order Psocoptera* (1,100 *species*)
These minute, fragile insects are closely related to the Zoraptera of the last chapter, and at one time the two were merged into one order. Yet in some ways the Psocoptera already begin to foreshadow the aphids, or plant-lice that are one of the culminating points of the hemipteroid line of evolution.

Psocoptera are scavengers on minute fragments of vegetable or animal matter, moulds, fungi, pollen, scraps of dead plants or animals, including dead insects. Those that live out of doors have no common name, and entomologists refer to them as 'psocids'. A number of species come indoors and scavenge in any dusty situation on similar fragmentary material, when they are known as 'book-lice'. They may be numerous, and seem to cause damage, for example to books, but it is not known for certain whether they really eat the paper, or only moulds that have already attacked it. They sometimes eat out the interior of insect specimens in collections, especially the juicier ones that have not become too desiccated.

Some psocids have well-developed wings, and are then recognised by the venation though many have the wings reduced or absent. They spin silk from a pair of glands opening on to the labium, and with the silk they make a net-like covering, under which a colony of psocids may assemble on the bark of a tree.

17. *Order Mallophaga* (2,600 *species*)

Many insects are called 'lice': this is one of those terms like 'bug' and 'fly' which have a restricted meaning in the technical sense and several colloquial ones. The term 'lice' is applied to minute, wingless, parasitic insects, which crawl about close to the skin of a warm-blooded animal and are difficult to dislodge. Mallophaga are 'biting lice' because they have biting mouthparts and scavenge among the debris of the skin, and are also called 'bird lice' because most of them live among the feathers of birds. Members of two families of Mallophaga, Gyropidae and Trichodelidae, feed on mammals, and infest domestic animals such as dogs, cats, horses, cattle, sheep and goats. Apparently none of them live on man, perhaps because he lacks a continuous coat of hair; though the little hair that man still retains may become infected with sucking lice of the order Siphunculata.

Mallophora are elongate, wingless insects, with relatively big heads and short, strong legs. These features are often developed in parasitic insects. Mallophaga normally feed on dead tissues, but will take blood from scratches or wounds. Perhaps through exploiting further this richer food supply the ancestral biting louse may have evolved towards the closely related order of sucking lice, Siphunculata.

The evolution of species in Mallophaga seems to follow to some extent the relationships of the host birds. Biting lice cannot long survive away from a host and, being wingless, they must be passed on by contact between two hosts. Hence they tend to evolve along with their host species with little chance of spreading to a host of a different species. This is in strong contrast to other, more mobile insects which parasitise birds: fleas, which live their larval life in the nest, and parasitic flies, which have wings. These insects are more commonly associated with the hosts that occur together in a particular environment, rather than with hosts that are zoologically closely related.

18. *Order Siphunculata* (*Anoplura*) (250 *species*)

The true, or sucking lice are also wingless parasites, an evolutionary development from the biting lice, but even more closely linked with their host, since they have developed piercing mouthparts, and live by sucking blood.

The best known member of the order Siphunculata is the Human Louse, *Pediculus* humanus, which provides an interesting demonstration of the way in which these insects have adapted themselves to their hosts. The Human Louse exists in two races, *P. humanus capitis*, the Head Louse, and *P. humanus corporis*, the

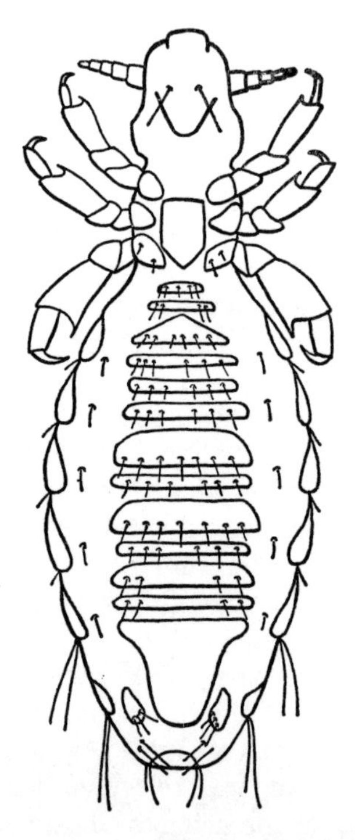

Figure 27. A sucking louse (order Anoplura=Siphunculata) (after Borror and Delong).

Body Louse. They are different in appearance, and were formerly thought to be different species, but it has been shown that *capitis* will turn into *corporis* if it is made to change its situation on the human body. It seems that race *capitis* is the primitive form of the species, adapted to living in a long coat of hair, which humans have lost and replaced by clothing. Race *capitis* persists in the hair of the head: the hair of the pubic region has evolved its own different species of louse *Phthirus pubis*, the Pubic Louse or Crab Louse. Meanwhile from the Head Louse, *P. humanis capitis*, has evolved a race adapted to living in clothing, a new and artificial habitat; this race is *P. humanus corporis*.

These lice lay large numbers of eggs ('nits'), which hatch into insects that undergo little or no metamorphosis. They have three moults in a life-cycle lasting about a month. They flourish when left undisturbed by combing, washing or the changing of clothing. Head lice are passed from one person to another by sharing combs or caps; body lice cling to the clothing when it is removed.

Sucking lice infest a wide range of mammals – but not birds – including such interesting victims are elephants and seals. The Pig Louse, *Haematopinus suis*, is the best known of these, and its big eggs attached to bristles of the pig are conspicuous objects.

Siphunculata carry many diseases, notably typhus and relapsing fever (see Chapter 19).

19. *Order Hemiptera* (55,000 *species*)

Though many insects are inaccurately called 'bugs', this name is correctly applied only to members of the order Hemiptera, all of which have abandoned the primitive mandibulate mouthparts, and have adapted their proboscis for piercing the tissues of plants, the cuticle of other insects, or the skin of warm-blooded animals. The Bed Bugs of the family Cimicidae, 'Bugs' *par excellence*, belong to this order of insects.

Hemiptera are not only the biggest order of Hemimetabolous insects, but also by far the most diverse, and so they may be looked upon as the crowning achievement of this line of evolution. They divide sharply into two series: Homoptera and Heteroptera. Some authorities reserve the name Hemiptera for the Heteroptera alone, and raise Homoptera to the status of a distinct order. From an evolutionary point of view it is probably more logical to regard them as divergent branches of one order. This view is supported by the existence in the ancient Gondwanaland countries of New Zealand, Australia and the tip of South America of a family called Peloridiidae, which in some ways links the two sub-orders. A detailed study of this transitional group is given by China [22].

Homoptera are known to have existed in Permian times, and their remains are very numerous in deposits of that period The more primitive families, many of which still exist today, are grouped as Auchenorhyncha, and from them are believed to have evolved, first the Heteroptera, and then the more advanced

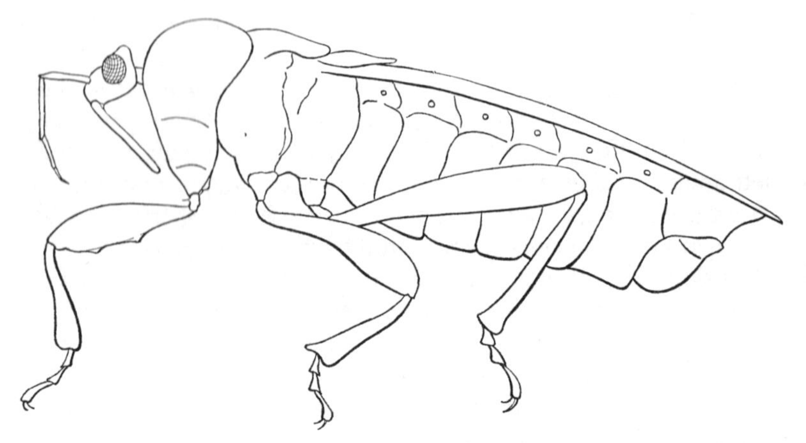

Figure 28. A bug (order Heteroptera) with mouthparts developed into a sucking proboscis.

families of Homoptera, called the Sternorhyncha. The fact that Heteroptera are not known as fossils before the Upper Trias accords with this view, though from the state that they had then reached it is considered that both Heteroptera and Sternorhynchous Homoptera must have separated off during the Permian epoch.

Peloridiidae have been classified as a third sub-order, Coleorhyncha, but China [22] prefers to regard them as 'living fossils' survivors of an early offshoot of Auchenorhyncha, which otherwise died out in Triassic times: a sort of entomological Coelacanth.

SUB-ORDER HOMOPTERA

All these are plant-feeding insects. A few of them are big, some Cicadas and some Lantern-flies (Fulgoridae) reaching as much as two inches, but most Homoptera are small, soft-bodied and yellowish or greenish in colour. When they have wings, both pairs are membranous, often with a complex venation of closed cells (e.g. a cicada, figure 29), but often the veins are few, and feebly chitinised. A great many Homoptera are wingless, at least in one sex, or for a number of successive generations, until winged generations reappear.

Homoptera fall into two groups of families, Auchenorhyncha and Sternorhyncha. Auchenorhyncha, the more primitive Homoptera, are the more mobile insects. Cicadas are famous for their sound production and ventriloquistic powers (see Chapter 9). Lantern-flies (Fulgoridae) are obese tropical insects which often have the front of the head inflated into a peculiar snout. They are called Lantern-flies because at one time this snout was thought, erroneously, to be luminous. To this group also belong the frog-hoppers, or cuckoo-spit insects (Cercopidae), which spend their nymphal life surrounded by a mass of 'spittle', coming out and moving about freely when they are adult; leaf-hoppers (Cicadellidae; Jassidae), more tubular insects; and tree-hoppers (Membracidae) which live mainly on trees and shrubs, and often have the thorax produced into horns or spines, which are believed to confer a protective resemblance to the bark on which they sit. Several smaller families occur in this group, and all except the cicadas are jumping insects.

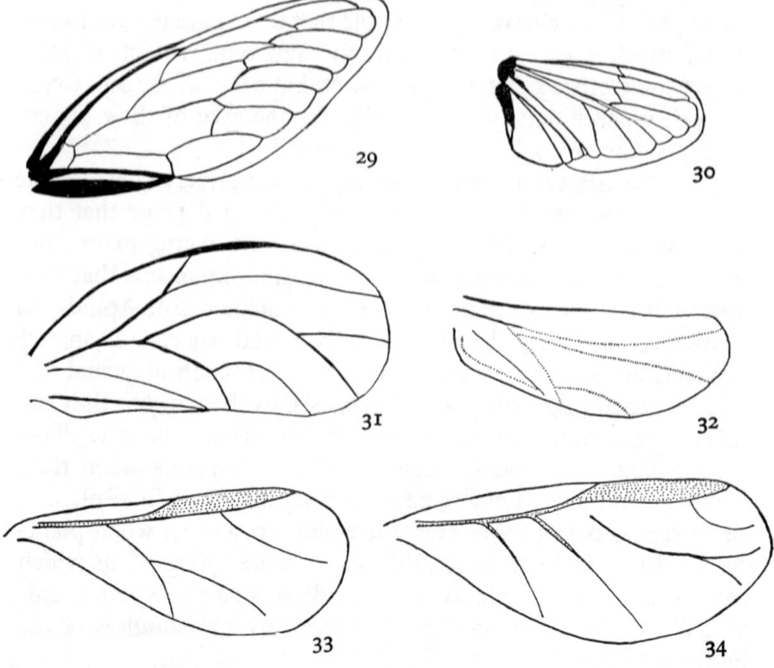

Figures 29–34. Wings of bugs of the suborder Homoptera: 29, 30, Cicadidae (fore- and hind-wings); 31, 32, Psyllidae (fore- and hind-wings); 33, Phylloxeridae (fore); 34, Aphididae (fore). Note that in spite of the name Homoptera, meaning that the wings are the same, the two wings of each side are not quite identical (after Borror & Delong).

Sternorhynchous Homoptera are the familiar Green-fly and Black-fly (Aphididae), White-fly (Aleurodidae), scale-insects and mealy-bugs (Coccidae) and jumping-plant-lice (Psyllidae). Except for the last named, members of this section are much less active than the preceding families, and live static lives in large colonies, feeding on various plants which they infest. The scale-insects are so called because the females are always wingless, and are protected by being covered with a shield, or scale, or a resin or a mealy substance. Underneath this the female feeds from the plant-tissue and is often without most of the usual external organs of more active insects.

Aphidae and Aleurodidae also produce wax, and cover the plant with a white incrustation. They also produce 'honey-dew', the surplus sugar from the sap upon which they have fed, which is voided with the faeces, and is a nuisance when it drips from trees on to pavements below. It is possible that these insects have to take in sap much in excess of their carbohydrate requirements in order to get some other substance such as amino-acids in the sap. Other insects, notably ants, pursue aphids for the sake of their honey-dew (see Chapter 17).

All these Sternorhynchous Homoptera are of very considerable importance as pests, not merely for the direct damage that they cause when they withdraw so much sap from young plants, but even more for the viruses and other harmful organisms that they spread from one plant to another (see Chapter 20). Aphids, in particular, are very highly specialised and successful insects biologically. and make much use of an alternation of sexual and asexual generations, with parthenogenesis. By these means they can increase in numbers very rapidly at certain seasons when the plant-food is abundant, and disperse to other food-plants when their summer food fails. They remain encrusted on their food-plant at these times, but at other seasons, mainly in winter, when plant-food is less readily available, they have winged generations which can fly off to other plants. Besides dispersing the species, this sexual reproduction revitalises the species by a re-shuffling of the hereditary material.

SUB-ORDER HETEROPTERA

It is to this sub-order that the name Hemiptera, or 'half-winged insects' is appropriate, because the fore-wings are half horny and half membranous. Such wings are called hemelytra, in comparison with the horny elytra, or wing-cases of beetles. Figure 35 gives the names of the various divisions of such a wing.

Heteroptera are a very successful and diversified group, and it is impossible in an elementary book such as this to attempt to discuss them. Various classifications have been proposed, but most authorities agree to divide them into water-bugs and land-bugs, distinguished not only by their habits, but by their antennal structure.

Cryptocerata – water-bugs – are highly adapted for life under

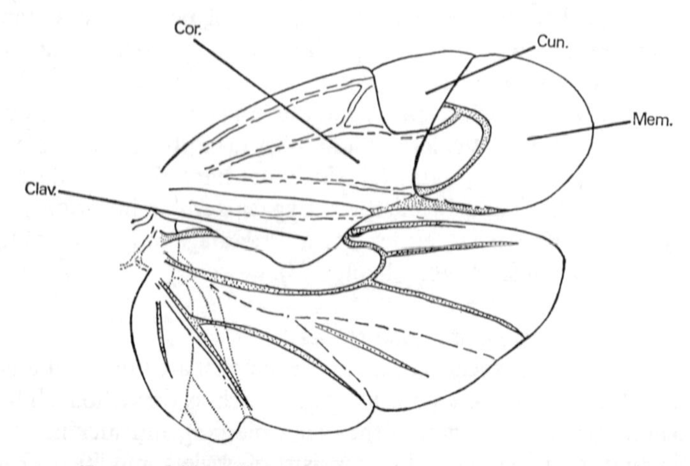

Figure 35. Fore- and hing-wings of a bug of the sub-order Heteroptera. Parts of fore-wing: clav = clavus; cor = corium; cun = cuneus; mem = membrane. (after Borror & Delong).

water both as nymphs and as adults. This is in contrast to May-flies or Dragon-flies which as we have seen, have an aquatic nymphal stage that is highly adapted for the water, followed by an aerial adult that has abandoned the water for ever.

The name Cryptocerata means 'with the antennae hidden', and this concealment is part of the streamlining for underwater movement that is continued in the smoothly shaped body and paddle-shaped limbs of Corixidae (water-boatmen), Notonectidae (back-swimmers), and Belostomatidae (the giant water-bugs of warm countries). All these swim actively, and carry with them a film of air, or *plastron*, applied closely to the under-surface of the abdomen, and acting as an underwater lung (see pages 128–131). In contrast, Nepidae (water-scorpions) and Naucoridae (creeping water-bugs) crawl about on submerged vegetation. Naucoridae have a plastron; but Nepidae have a respiratory tube from which they get their name 'water-scorpion', and some authorities think that they are an offshoot of the Gymnocerata, close to the predatory Reduviidae. The true water-bugs are carnivorous, with the exception of the large family Corixidae which includes many algal-feeding species.

The group Gymnocerata – land-bugs – also includes the pond-skaters and water-striders (Hydrometridae), those familiar spider-like insects that run about on the surface of ponds and streams without getting wet (plate 35). They are terrestrial group that has acquired the habit of scavenging for dead insects trapped on the water. The Gymnocerata as a whole are a terrestrial group, and include some of the largest and most brightly coloured plant-feeding bugs. Pentatomidae (shield-bugs) have the scutellum of the thorax conspicuously enlarged. Lygaeidae, Pyrrhocoridae and Tingidae are often pests, either by direct damage or because of the diseases they transmit (see Chapter 20).

20. *Order Thysanoptera* (2,000 *species*)

These are the tiny insects known as 'thrips' that are often abundant on the flower-heads of composite plants such as dandelions. They also attack every other part of the plant, piercing and sucking with their tubular proboscis: this consists of stylets modified from maxillae and mandibles, but is curiously asymmetrical, the right mandible being small or vestigial. Like aphids, thrips may damage plants just by removing a great quantity of sap, but they aggravate this damage by transmitting diseases from one plant to another.

The name Thysanoptera refers to the fringed wings, which are unique among insects. The true wing is reduced to a narrow strap, which is fringed with long hairs before and behind. The two wings on each side are locked together at the base by a series of hooks, and are folded over the back when at rest. Thysanoptera are divided into two sub-orders: Terebrantia, with a saw-like ovipositor, lay their eggs inside the tissues of plants; Tubulifera have, at most, a simple tubular ovipositor, and deposit their eggs in crevices. The pests of crops mostly belong to the Tubulifera.

From an evolutionary point of view, Thysanoptera are interest-

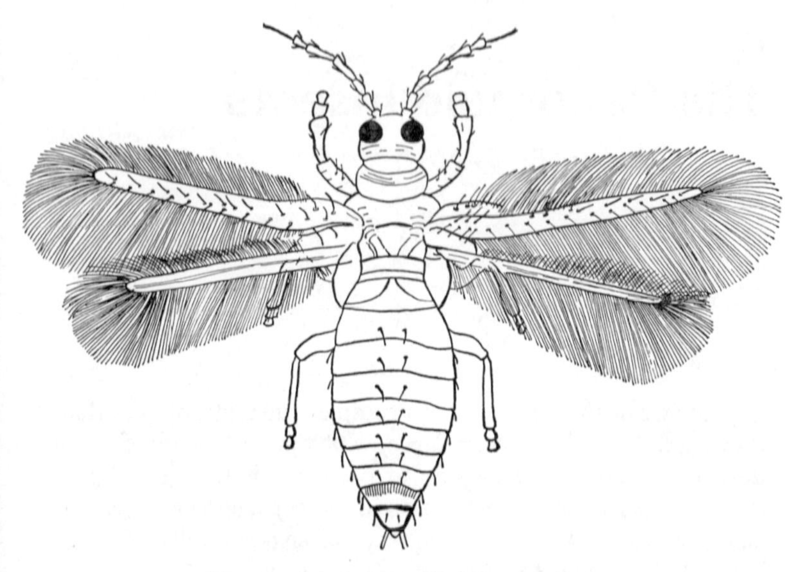

Figure 36. A thrips, order Thysanoptera.

ing because they have a curious metamorphosis. Terebrantia have four nymphal instars, and Tubulifera have five. In each group, the first two instars are like the adults, except that they have no external signs of wings. They move actively and feed, but the remaining instars are inactive and do not feed, and may be enclosed in a cocoon or an earthen cell. The final instar before the adult is called a 'pupa' and the one or two preceding inactive stages are called 'prepupae'. In both sub-orders the last two instars have large external wing-pads.

This life-history has been considered to be in some ways intermediate between those typical of Hemimetabola and Holometabola, as is suggested by the names 'prepupa' and 'pupa' for the inactive instars. Except for the loss of external wing-pads, the first two instars of Thysanoptera have not become different from the adult, so this mechanism, if indeed it is the same as that of the Holometabola, must be fixed in a very rudimentary stage.

6
The Panorpoid insects

All the rest of the insects have a complete metamorphosis; that is they hatch from the egg as a *larva*, which pursues a life of its own until it is fully fed and ready to become adult. Then it passes through a period of reconstruction in the *pupa*, and emerges as the *imago*, or adult insect. The sequence of egg-caterpillar-chrysalis-butterfly is familiar to everyone.

Surprisingly, perhaps, this useful device seems to have evolved only once. Though the details vary, the basic plan is the same in all holometabolous insects, and it is generally assumed that they all came originally from one ancestral form. If there is any doubt, it would be about beetles, which, as we shall see, stand apart from all the others except for a peculiar group of parasitic insects, Strepsiptera. Except for Thysanoptera there are no intermediates between being hemimetabolous and being holometabolous, and perhaps this in itself implies that the decisive step was taken long ago.

As is often the case among the more highly evolved groups of animals, the orders in this section of the Insecta are fewer and bigger, and they include the Big Four of the Insect World: Lepidoptera (butterflies and moths), Diptera (true flies) Hymenoptera (bees, wasps and ants) and Coleoptera (beetles). An immense amount is known about these four orders, each of which contains insects of a wide range of sizes, habits and life-histories. Just because there is so much detailed information about them, these orders cannot be adequately discussed in an elementary book like the present. Each of them is covered by textbooks and by specialist works, to which the reader is referred.

Super-order Panorpoidea

Five of the nine orders of holometabolous insects form a natural group, which Tillyard called the *Panorpoid Complex*, but which is now more generally called the super-order Panorpoidea, comparable with the Orthopteroidea and Hemipteroidea. *Panorpa* is a genus of Scorpion-flies, common insects of the hedgerow, and the order Mecoptera to which this genus belongs appears to be a focal point of this group of orders. We may add to it the transitional order Neuroptera at the start of the series.

21. *Order Neuroptera* (5,000 *species*)

Fragile insects, with two pairs of membranous wings, often with a complex venation including a network of crossveins, and with extra – 'supernumerary' or 'intercalary' – veins, Neuroptera have no common name. Familiar examples of the order are the green lacewings (Chrysopidae) that often come indoors at night; the brown lacewings (Hemerobiidae) and the alder-flies (Sialidae) that make furtive flights near water. Such insects have superficial resemblances to more primitive aquatic insects, such as the Stone-flies (Plecoptera) which are now placed near the stem of all living Hemimetabola. Neuroptera are Holometabola, but have the same appearance of arising from the base of the evolutionary stem. It has also been suggested that Psocoptera, the most primitive order of Hemipteroidea are also closely related to the Hemerobiidae, a family of Neuroptera. Neuroptera are a rather varied assortment of insects, and are often divided into two orders, Megaloptera and Planipennia.

Megaloptera are the Alder-flies and the Snake-flies, and have carnivorous larvae with chewing mouthparts. The larvae of *Sialis* and of *Corydalis*, the large Dobson-fly, live under water, and feed upon other insect larvae, crustaceans and worms (plate 37). Like the nymphs of May-flies and Stone-flies, the larvae of Neuroptera have tracheal gills along the sides of the abdomen. The adults are sluggish insects, more heavily built and slow flying than the lace-wings. These waterside insects are called Alder-flies (figure 64).

The related Snake-flies (*Raphidia*) (figure 63) are terrestrial, and their larvae feed upon other small insects, which they catch under bark. Snake-flies are easily recognised by the elongate prothorax, which holds the head forward on a long 'neck', and makes these insects look like a miniature of one of the great reptiles.

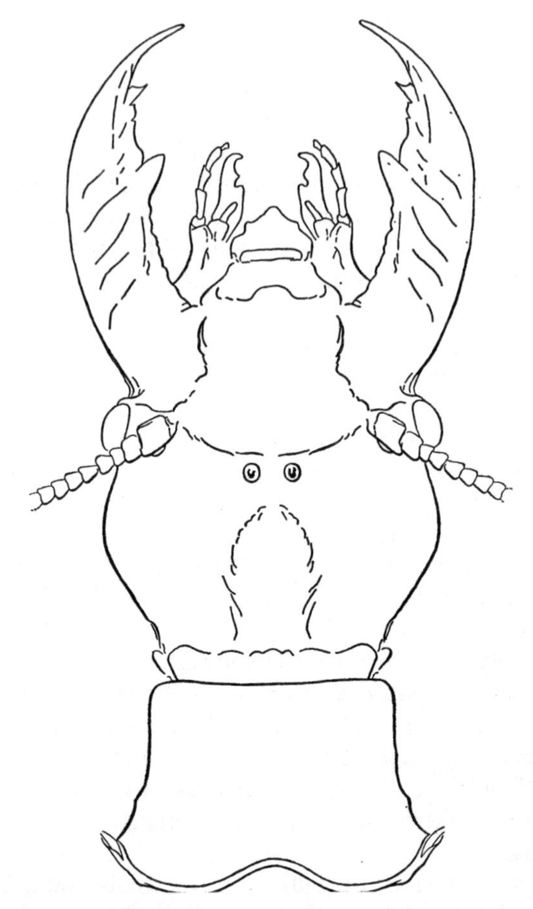

Figure 37. Head of a dobson-fly (order Neuroptera).

Planipennia, again, are a mixed group. The fragile, delicate lacewings, with their large and sometimes beautifully golden eyes (hence the name *Chrysopa*) have already been mentioned. Others in the groups are Mantispidae, with the fore-legs modified for grasping, like those of a praying mantis; the ant-lions (Myrmeleontidae); and other families such as Ascalaphidae and Nemopteridae, none of which have a common name. Nemopteridae are famous for the way in which the hind-wings are drawn out into long, paddle-like streamers. The fore-wings are used for active flight, and the hind-wings trail behind.

The larvae of all Planipennia are carnivorous, but have sucking mouthparts, an evolutionary advance from the chewing types of the Megaloptera. The mandible and maxillae on each side are curved and fit together to form a pair of piercing and sucking fangs. Larvae of many (but not all) of the ant-lions live in pits in sand or dust and devour ants and any other small creatures that may fall into the pits – 'Demons of the Dust', Wheeler called them. Larvae of Mantispidae prey within the egg-cases of Lycosid spiders; those of lacewings feed on aphids; those of Osmylidae prey on other aquatic insect larvae and have no tracheal gills, whereas those of Sisyridae, which live in and upon fresh-water sponges, have gills. The only generalisation one can make about the biology of Planipennia is that they are a vigorous and versatile group, and attractive to the field student (plate 24).

22. *Order Mecoptera* (350 *species*)

Although this is a small order in the number of species now living, it is a key group in understanding the relationships and probable evolution of the most advanced insects.

Panorpa, the Scorpion-fly, is well named as figure 38 shows. Though there are not many species, individuals are common in most temperate countries. They sit about on damp vegetation, and in their short flights look like something between a moth and a Crane-fly. The head is drawn out into a snout – again as in Crane-flies – and the fore- and hind-wings are very much alike.

Mecoptera have a complete metamorphosis, and the larva is eruciform: that is, it is like a caterpillar, except that it has eight pairs of abdominal prolegs, including one pair on the first segment. Adults as well as larvae are mandibulate, and chew animal food, though it is not certain whether they eat living insects or only dead ones.

Usually included among Mecoptera are the Boreidae,* the Snow Scorpion-flies, which live in moss in cold situations, and may be

* Hinton [43] has proposed that the Boreidae should be removed to a separate order, Neomecoptera, as the most primitive section of the Panorpoidea. He further accepts the order Zeugloptera for the Micropterygidae, the most primitive family of Lepidoptera, and then arranges the six principal orders of Panorpoidea into two groups of three: Zeugloptera, Trichoptera and Lepidoptera in one; Siphonaptera, Mecoptera and Diptera in the other. He does not agree with the usual view that the fleas (Siphonaptera) may be an offshoot of the flies (Diptera), but derives them directly from a primitive Boreid-like ancestor.

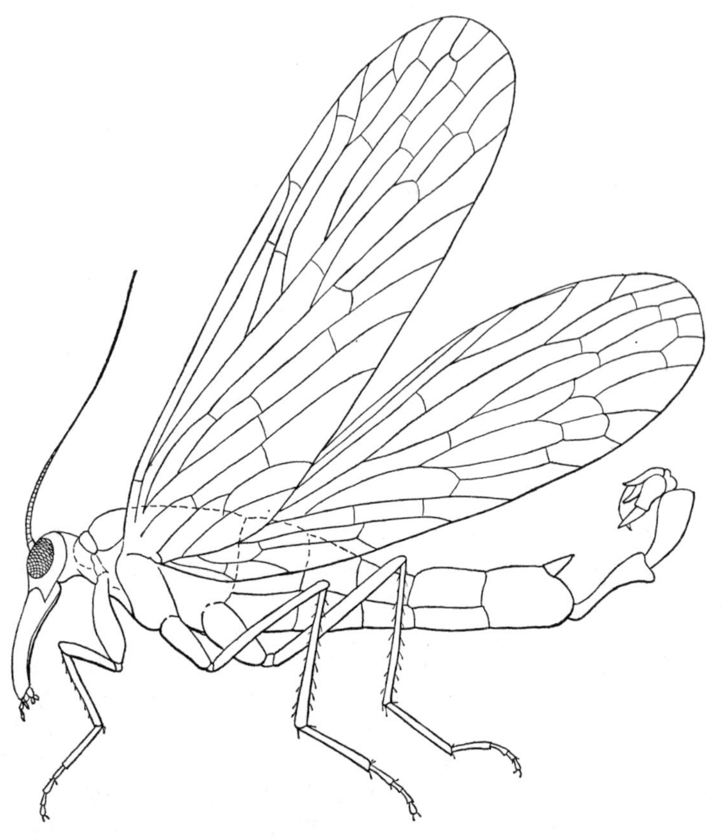

Figure 38. A scorpion-fly (order Mecoptera) (after Borror & Delong).

found on snow; and Bittacidae, the Hanging Scorpion-flies, which dangle from twigs and catch other insects. Both families have members that are short-winged, or entirely wingless, and have several features in common with some Crane-flies.

23. *Order Trichoptera* (4,500 *species*)

Caddis-flies are well-known to fishermen because their aquatic larvae are an important food of fish. The larvae are like small caterpillars, but as an adaptation to aquatic life the abdomen is thinly chitinised, and usually has filamentous tracheal gills. Most caddis larvae – inaccurately called 'caddis-worms' – live in cases, which they make from a framework of silk to which adhere fragments of gravel, small stones, etc. (figure 39). Nearly all caddis larvae live in fresh water, but there are a few salt-water forms, and some live among moss. One, *Enoicyla pusilla*, lives in leaf-litter, and so is as fully terrestrial as most insects (see Drift, 1957).

Adult Trichoptera have wings that look superficially like those

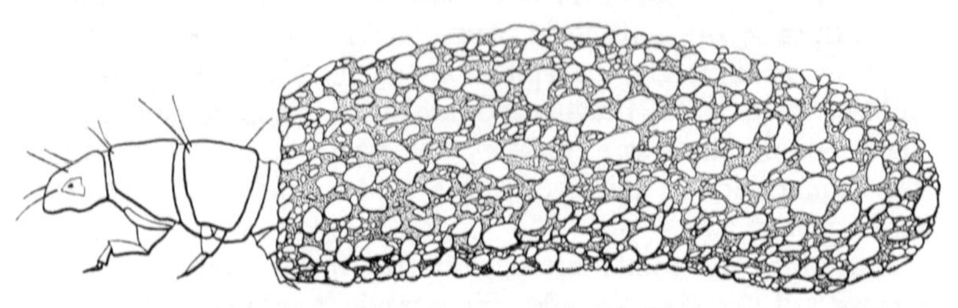

Figure 39. Larva of a caddis-fly (order Trichoptera) living in an underwater case made from silk and small stones (after Ross).

of moths, but are covered with hairs instead of scales. They have sucking mouthparts, but do not obviously feed, and are sluggish, furtive insects. Their relationships with Lepidoptera is shown by the caterpillar-like larvae, by the wings and proboscis of the adult.

24. *Order Lepidoptera* (200,000 *species* (plates 8–10, 25, 31, 39)

Everyone knows butterflies and moths. As adults they have large, scaly wings, supporting a relatively puny body, and long antennae, clubbed in butterflies, and with a great variety of shapes in moths. The larvae of Lepidoptera are the familiar *caterpillars*, with a hard, well-formed head, three pairs of short but strong thoracic legs, and with prolegs on at least some of the abdominal segments – usually on five segments, beginning on the third. The prolegs have a circlet of hooks, or crochets, and they are actively used in conjunction with the thoracic legs, as part of the normal walking mechanism.

This order, like the two following ones, is too big and diverse to be discussed in a few words. There are very many books about butterflies, and these should be consulted for information about the classification, biology and habits.

It can briefly be said that adult Lepidoptera nearly all have a sucking proboscis, derived from the galeae of the maxillae (plate 39), and that they feed on nectar, pollen, sweat and liquids from dung and decomposing materials in general. The impression that these beautiful insects take only ambrosial food is a romantic fiction. Adults of the primitive family Micropterygidae have mandibulate mouthparts, and mainly for this reason some authors place them in a separate order Zeugloptera (see footnote, page 65).

Larval Lepidoptera, caterpillars, are mandibulate, and mostly chew leaves, or mine in them, or bore into stems or fruit. Sometimes their activities result in plant-galls. A few larvae are predaceous on other insects. Caterpillars are among the most destructive of insect pests, partly because they are prone to appear suddenly in enormous numbers as a 'plague', but mainly because each larva eats so much and so quickly, that the sudden defoliation is a shock to the plant.

Butterflies and moths are superficially much more alike than members of any other big order, for example Hemiptera or Hymenoptera. The layman's idea of an entomologist is of a 'butterfly hunter', who shows you row upon row of pinned and set Lepidoptera, and enthuses about trivial differences in the pattern of the wings. This has the unfortunate effect of encouraging serious students to dismiss them as homogeneous, and inferior in interest to the more diverse orders. In fact, Lepidoptera are an advanced and highly successful order, whose detailed life-histories provide a wealth of material for the study of many aspects of evolution and adaptation, including much fundamental work on genetics and heredity.

25. *Order Diptera* (85,000 *species*) (figures 5 & 72 and plate 48)
These are the 'true' flies in the entomological sense, omitting those many insects of other orders, Dragon-flies, May-flies, Caddis-flies and so on, which we have already discussed, and a few more, such as Ichneumon-flies, which belong to the Hymenoptera, and Fire-flies, which are beetles. True flies have only one pair of functional wings, the fore-wings, placed on the mesothorax. The wings of the metathorax have become reduced in size and modified in shape, and are the stalked balancing organs known as *halteres*.

This is a very large order, of which a detailed account is beyond the scope of this book. Flies are grouped into three sub-orders: Nematocera, Brachycera and Cyclorrhapha. Primitive flies of the Nematocera are mostly elongate and fragile, the Crane-flies (daddy-long-legs) being an extreme example. The trend of evolution has been towards short, compact, often bristly flies, with short, broad wings, culminating in such flies as the House-fly, and the Bluebottle. The sub-order Cyclorrhapha includes these and many others, such as the famous fruit fly *Drosophila*. The intermediate sub-order Brachycera includes a miscellaneous group of some of the biggest flies, including the Horse-flies (Tabanidae) and the Robber-flies (Asilidae).

All adult flies have sucking mouthparts, somewhat similar in principle to those of the bugs (Hemiptera, Chapter 6). They may suck either through a tube made from the elongate mandibles and maxillae or through the labium, which has sponge-like lobes called *labella* at its tip; or by a combination of the two. Among Nematocera and Brachycera there are many well-known bloodsucking groups such as the mosquitoes, and in these only the adult females can pierce the skin of vertebrate animals and draw blood. Cyclorrhapha have lost the elongate mandibles and maxillae, and with them this method of bloodsucking. Most of them, like the House fly, mop up liquid foods with the labella, but some, notably the Tsetse-fly (*Glossina*) and the related Stable-fly (*Stomoxys*), have hardened the labium, and use it to pierce and draw blood. When this development has occurred it is found in both sexes.

The feeding of adult Diptera is supplementary to that of the larva, which is the principal feeding-stage. Larvae of flies have no thoracic legs, though they often develop prolegs. These are most elaborate in aquatic larvae, which crawl about on stones. Terrestrial dipterous larvae are mostly grubs or maggots. Generally the larva of the more primitive families have a well-developed head, and live by scavenging among animal or vegetable debris: they comprise the aquatic forms such as mosquitoes, gnats and aquatic midges, and the terrestrial forms such as the fungus-gnats, which live in the soil or in compost-like decaying vegetation. Cyclorrhapha have developed the *maggot*, pointed at the head, and blunt posteriorly, with a prominent pair of rear spiracles; a universal type, that can adapt itself physiologically to live in any kind of soft medium, animal or vegetable, living or dead.

26. *Order Siphonaptera* (*about* 1,000 *species*)

Fleas are a homogeneous group, a curious sideline of the Panorpoid stem. They are usually said to be closely related to the Diptera, but Hinton [43] derived them directly from a Neomecopteran ancestor (see footnote, page 65). They are wingless, and flattened from side to side, with tough, leathery bodies, and powerful hind-legs adapted for jumping. They are fully adapted to living as external parasites of mammals and birds.

Adult fleas have sucking mouthparts and feed only on blood. Unlike ectoparasitic flies, such as Hippoboscidae (e.g. the Sheep Ked, *Melophagus ovinus*), fleas have not become viviparous, and still have an independent larva stage. They lay eggs in the fur or feathers of their host, and the eggs or the young larvae then fall off into the surroundings. Flea larvae scavenge among debris, especially of human origin, fragments of skin or feathers, dried blood and so on, which may have fallen from their host. Hence fleas are really attracted not so much to their host as to their host's dwelling-place.

Fleas will usually bite any warm-blooded animal they may encounter, including man, but they can set up a breeding population only in association with their specific host, or one of closely similar living habits, because it is to the breeding-site that the fleas are specifically adapted. Fleas are therefore parasites of mammals or birds that have a nest or lair to which they return from time to time; they do not infest nomadic animals, such as horses or cattle, except as strays.

Like other bloodsucking insects, fleas are liable to carry disease, and are specifically linked with bubonic plague and with myxomatosis of rabbits (see Chapter 19).

7
Hymenoptera, Coleoptera and Strepsiptera

Three remaining orders require special consideration.

27. *Order Hymenoptera* (*perhaps* 100,000 *species*)
The bees, the wasps, the ants, the saw-flies and a great variety of parasitic insects are linked together in this order, which is called Hymenoptera in reference to the membranous wings. The wings are very highly developed, and their efficiency is increased by having the two wings of each side secured together with a row of hooks. The wings of many Hymenoptera have a characteristic arrangement of the veins into closed 'cells' (figure 40, plate 26), the hind margin of the wing remaining flexible. In the smallest Hymenoptera the veins are reduced in number, or almost completely lost, as in the Chalcid shown in figure 41.

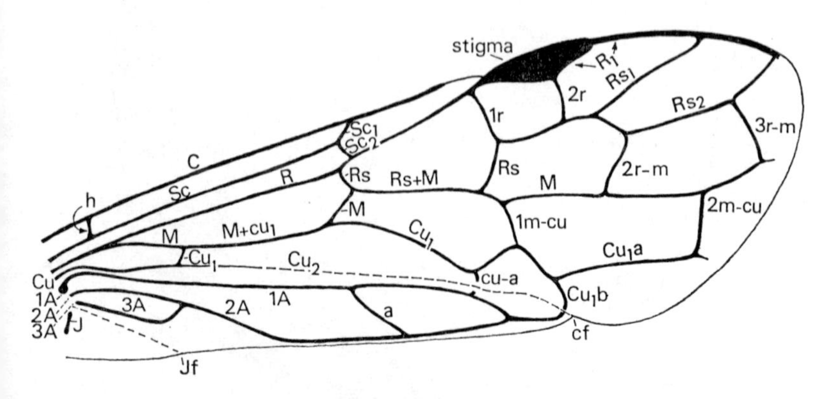

Figure 40. A generalised diagram of a hymenopterous wing, with nomenclature of veins (after Ross).

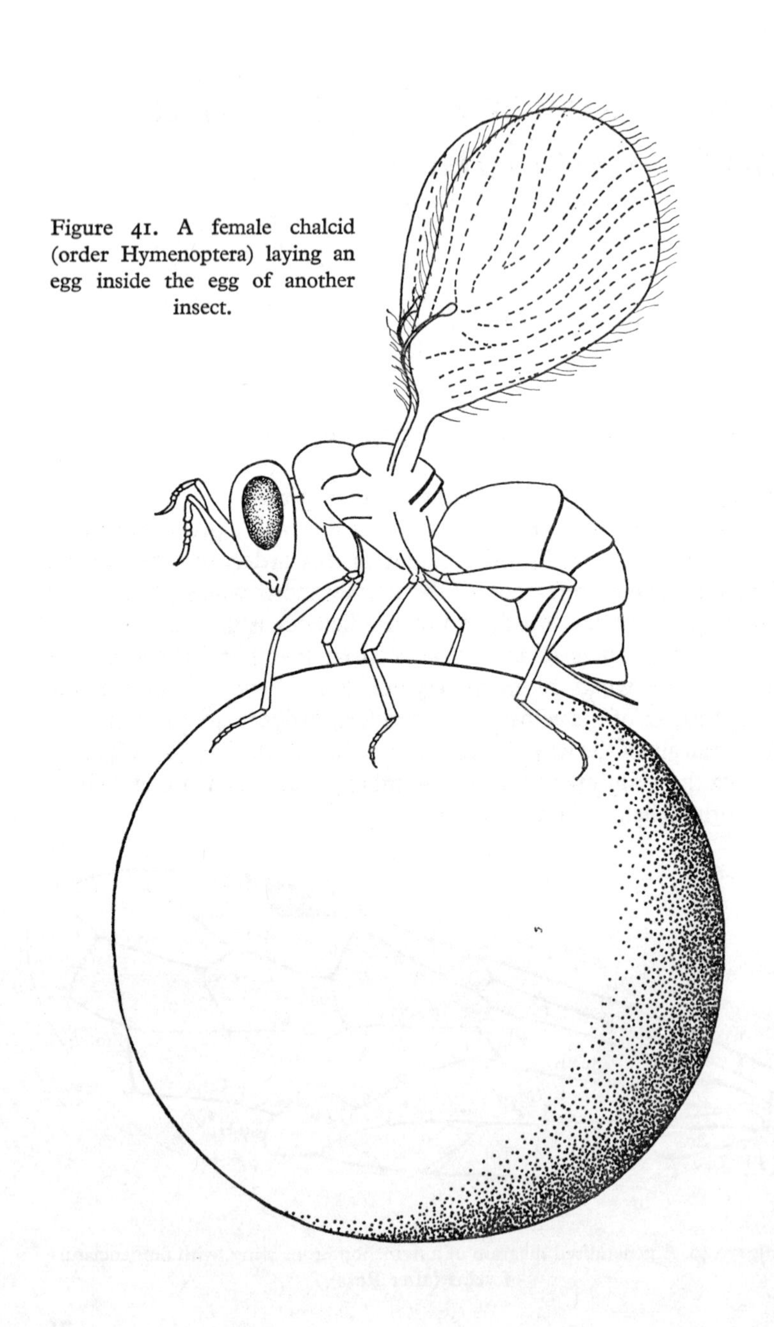

Figure 41. A female chalcid (order Hymenoptera) laying an egg inside the egg of another insect.

This characteristic wing-venation is among a number of characters which make it difficult to relate the Hymenoptera closely to any other order. Like Lepidoptera and Diptera, Hymenoptera are of comparatively recent origin, and seem to have evolved along with the flowering plants in Tertiary times. Because of their diversity it is necessary to give a little more space to them than to the other big orders.

SUB-ORDER SYMPHYTA

The most primitive Hymenoptera are the *Saw-flies*, distinguished from the rest by having no 'wasp-waist'. Female Saw-flies have a well-developed ovipositor with which they insert their eggs into the tissues of a plant, or into wood: Tenthredinidae have this ovipositor saw-like and are the true 'Saw-flies'; among the others are the 'wood-wasps' (Xiphidridae), 'horntails' (Siricidae) and the Stem-saw-flies of cereal crops (Cephidae). Larvae of Saw-flies still have thoracic legs, and those of them that feed externally on foliage also have abdominal prolegs like a caterpillar, but more of them, including one pair on the second abdominal segment (true caterpillars never start before the third segment). All Saw-fly larvae are plant-feeders, except those of Orussidae, which are parasitic on the wood-boring larvae of Buptrestid beetles. Saw-flies can be very destructive, whether of trees or of herbaceous plants. Some larvae are leaf-miners or gall-makers.

SUB-ORDER APOCRITA

These are the 'waisted' Hymenoptera, often divided further into *Aculeata* – the stinging forms, bees, wasps and ants – and *Parasitica* – the Ichneumon-flies, chalcids, and a host of others, generally small in size, the larvae of which are parasitic, This distinction is illogical in practice, and it is better to think of the Apocrita as falling directly into a number of super-families, Vespoidea, Apoidea, Formicoidea, Ichneumonoidea, Chalcidoidea, Proctotrupoidea and so on (Table 3).

Larvae of these Hymenoptera are mostly pale, legless and feeble, grublike creatures that can only survive in an abundance of food. This state has evolved along with the development of some degree of parental care, even if this is no more than a careful selection by the female of the precise spot in which to lay her egg or eggs. The eggs of the parasitic families are laid on or in the egg, larva or pupa of another insect, or of another arthropod, and very pretty studies have been made of the way in which the female recognises a suitable victim. Even after correct selection there is a great risk of wastage among the eggs and larvae, and survival is improved by such biological devices as *parthenogenesis* and *polyembryony* (i.e. the development of a number of larvae from one egg. Many of these insects are *hyperparasites*, living at the expense of other larvae – Hymenoptera or not – which are themselves already parasites of the primary host. Because they kill many other insects, and may choose their prey quite specifically, parasitic Hymenoptera are the most important group of insects involved in *biological control* (see Chapter 22).

Cynipoidea, the gall-wasps, are a strange offshoot of the Apocrita, with larvae that are plant-feeding. The larvae live internally, and cause the plant tissue to divide, and then to form a visible growth known as a plant-gall. This is really a form of plant-parasitism, and related families grouped together within the sub-order Cynipoidea may be either gall-forming or parasitic on insects.

The 'stinging' Hymenoptera are so-called because they have adapted the ovipositor into a weapon with which they can pierce a victim and inject a paralysing venom. This is readily used in defence, but its primary purpose is to paralyse the prey upon which the more primitive, solitary members of this group feed their larvae.

Sphecoidea and Vespoidea are wasps, and show all stages of evolution between solitary and social habits. The adults feed on carbohydrate food, nectar, sap, sugar, etc., but they give their larvae a protein diet by supplying them with animal food. The solitary wasps make a nest or cell in a hole, or from mud, sting some insect or other small Arthropod, bring it to the nest, lay an egg in or on it, and then seal it up. The newly hatched larva thus finds itself provided with food, but without maternal attention. The social wasps, on the other hand, live in colonies of varying size, and tend their larvae, bring them food mainly of animal origin.

Apoidea are the bees, which have given up a carnivorous diet, and feed themselves and their larvae on pollen and nectar. *Pollen* the male germ cells from plants, provides the protein. *Nectar* is the sugary substance produced by insect-pollinated plants, which acts as a bait, inducing insects to enter the flower-head and thus brush against the pollen, carrying some of it to the female organs of the same plant or of another one. *Honey* is the nectar regurgitated from the crop or 'honey stomach' of the worker bee, with the sugars partly hydrolysed, or 'inverted'.

Social insects are discussed more fully in Chapters 15–17. Among both wasps and bees the division between solitary and social mem-

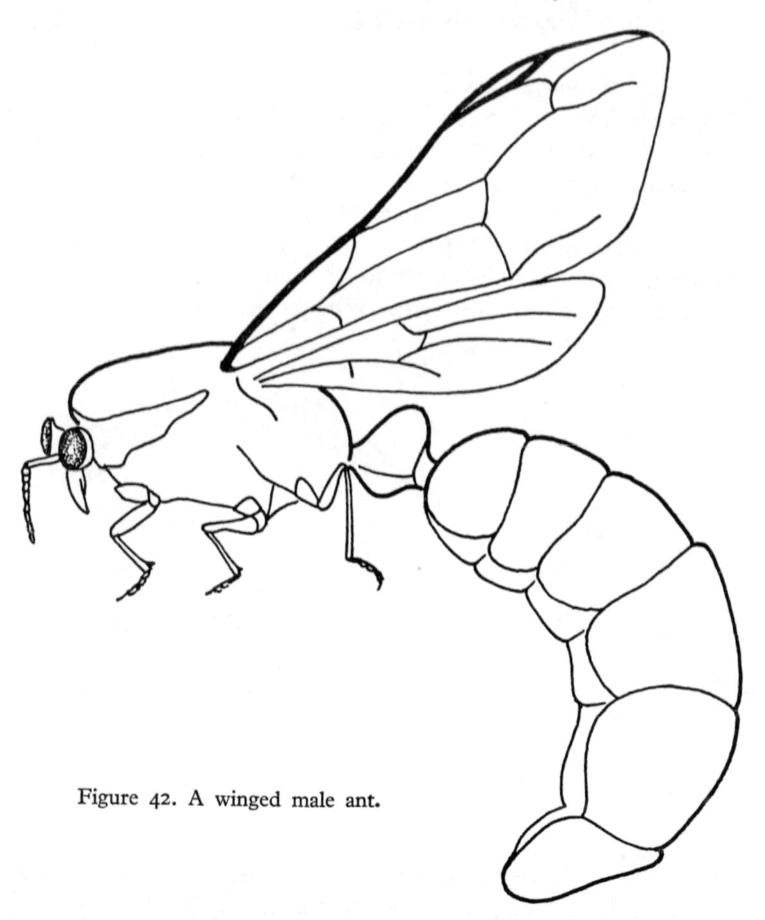

Figure 42. A winged male ant.

bers is not as sharp as may have been implied. The fully social groups are very much in the minority, and among the others there is a wide range of habits and behaviour. Bees can generally be distinguished from wasps by being stouter and more furry in appearance, the hairs of head and thorax often being feathery and producing the appearance of down.

Ants are the most highly evolved Hymenoptera, and they have carried the evolution of behaviour further than the wasps or the bees. Primitive ants are carnivorous, but most groups will eat almost anything. Moreover they have developed the art of *trophallaxis*, or mutual exchange of food with their own larvae or with guests such as asphids (plate 47).

28. *Order Coleoptera* (275,000 *species*)

Beetles are familiar insects to everyone, with their characteristically armoured appearance, the sclerites of the body usually heavily chitinised and precisely articulated. Most beetles have two pairs of wings, but the fore-wings are modified into wing-cases or *elytra*. At rest these cover the folded hind-wings, and usually conceal the abdomen, except in a few exceptional families such as Staphylinidae and Silphidae. The elytra seem to represent an evolution

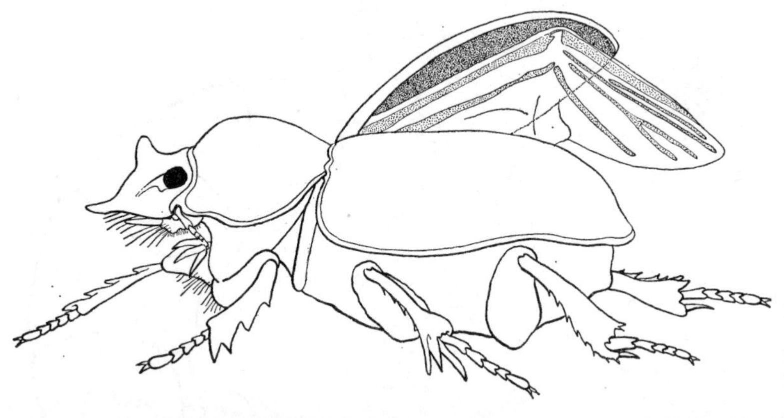

Figure 43. A stag-beetle (order Coleoptera), showing relation of hind-wing to fore-wing (elytron) (after Ross).

beyond that of the tegmina, or horny fore-wings of the Orthoptera. The tegmina of Orthoptera take part in flight, but the elytra of beetles are held aloft (plate 32). To some extent they act as fixed planes, increasing the lift, but their main function is to protect the hind-wings. The latter are large in beetles, and elaborately pleated.

Beetles do not belong to the Panorpoid complex and have no obviously close relationship to any other order. They first appear in deposits of the Permian period, about the same time that the Panorpoidea were being launched with primitive Neuroptera and Mecoptera. As an order they are full of contradictions. They have been immensely successful, and it is thought that there may eventually prove to be as many different species of beetles as of all other insects combined. They range in size from tiny to huge; the goliath beetles of the tropics are both bulky and heavy, and some *Dynastes* may be over one hundred and fifty millimetres long. Yet in spite of this, and of an endless diversity in details, which makes them one of the most popular orders to study, they give a general impression of uniformity to the casual observer.

Again, in spite of their long evolution, most beetles have remained terrestrial, ground-living, running insects. One searches for beetles under stones, in moss, in soil, round the roots of plants, in tunnels in wood, and in all such places where a furtive insect can take cover. We think of flies or Dragon-flies as aerial insects, and if we see one at rest we think of it as having 'settled'; in contrast beetles on foot seem in their natural setting, and their flight – accomplished though it may be – seems an occasional activity.

There are two main groups of beetles, a primitive group called Adephaga and a much larger and more varied group called Polyphaga. The first group are carnivorous both as adults and as larvae, and include the voracious tiger-beetles (Cicindelidae) and ground beetles (Carabidae), and aquatic carnivores such as the formidable Dytiscidae and the peculiar whirligig beetles (Gyrinidae) (plate 36).

Polyphaga, as their name implies, eat a great variety of different substances. It is inevitable that much of what they eat should be important to man, and so this group includes all the well-known pests: the beetles that feed in stored products of all kinds, in grassland, and of course the pests of timber, living and dead. It is hopeless to attempt to discuss beetles in the space available, and one of the excellent textbooks should be consulted.

29. *Order Strepsiptera* (300 *species*)

This survey of the orders of insects ends with one that is probably the most peculiar of all and which has no common name. Except for one family, Mengeidae, which live under stones, the females of all Strepsiptera are wingless and often live as parasites of adult insects of other orders. The best known is the genus *Stylops*, of which the female parasitises bees; others attack Homoptera, Heteroptera, Thysanura and Orthoptera. Male Strepsiptera (stylopids) have huge, fan-like hind-wings, and the fore-wings are reduced to halteres: the reverse of what has happened in Diptera.

The female remains between the segments of the host, usually those of the abdomen, and is fertilised in this position by the winged male. Very large numbers of tiny larvae, called *triangulins*, escape from the female and migrate in search of another host. Having found a host the larva moults and becomes a legless grub, which feeds in the host's tissues, pupating inside the last larval skin. The Strepsipteron emerges as an adult, to fly away if it is a male. A female remains in situ, awaiting fertilisation and the production of a new batch of larvae.

Strepsiptera are usually placed next to the Coleoptera, but some authorities think that they have links with Hymenoptera.

8
Vision in insects

Most adult insects have eyes, which they appear to use in much the same way as we use ours. In their daily life they behave as if they could see where they were going, especially in flight, when they avoid obstacles, hover over flowers, and alight safely, skilfully making allowance for the drifting effect of the wind.

Types of eye

A winged insect – that is an adult member of the sub-class Pterygota – often has eyes of two kinds: a pair of *compound eyes* (figures 3 & 41), one on each side of the head, and a set of three simple eyes, or *dorsal ocelli* (plate 27) arranged in a triangle at the top of the head, or vertex. Not all insects have the complete set. It is common for a species to have no ocelli, less common to have two or only one. Compound eyes also may be reduced in size or lost completely, as we shall see presently.

Nymphs of hemitabolous insects may have the same optical equipment as their parents, but larvae of Holometabola never have compound eyes. If they have visible eyes these are of a third kind known as *lateral ocelli*, or *stemmata*.

Compound eyes

These are the eyes proper, the principal organs of vision of adult insects. As such they are usually conspicuous, and generally much bigger in relation to the whole animal than are the eyes of mammals. Indeed in many aerial insects such as Dragon-flies (Odonata) and true flies (Diptera) they occupy by far the largest part of the volume of the head, and the other essential structures, mouthparts, pharynx, even brain, are crowded into a small space (plates 37 & 39).

On the other hand the eyes are generally smaller in ground-living insects such as cockroaches and beetles, and in sedentary feeders such as the sap-feeding Homoptera. They tend to dwindle or to disappear altogether in insects, belonging to any order, which have adapted themselves to living permanently in caves or other dark places, or as parasites in other animals. Thus worker ants usually have smaller eyes than the winged males, which have to find a mate in the open air; and by convergent evolution many of the inquilines, or insects of other orders that live in the nests of ants (or of termites) have small eyes or none at all.

The size and complexity of the compound eyes gives a clear indication of the sort of life that is led by the insect concerned and it is remarkable how quickly – on an evolutionary scale of time – the eyes become modified to suit the life of the insect. The order Strepsiptera are a good example of this. The females of most Strepsiptera, including the best-known genus *Stylops*, are wingless parasites, highly specialised and without eyes. The corresponding males have the eyes fully developed, though not excessively large. The most primitive family, Mengeidae, have females that are less specialised, and in some genera such as *Mengenilla* the females are not parasitic, live freely under stones, and have normally developed eyes. Since Strepsiptera are best known as parasites of bees and wasps, a recent group that has evolved since the flowering plants, the evolution of Strepsiptera is probably also a recent one, though some members of the order do parasitise more primitive insects, Homoptera, Orthoptera and even Thysanura.

Structure of the compound eyes

The essential difference between a compound eye and an ocellus is that the compound eye is a bundle of optical units complete in themselves, whereas in an ocellus, as we shall see later, one lens covers a complex of light-sensitive cells. Each unit of a compound eye is called an *ommatidium*, and has its own lens formed from an area of transparent cuticle. The surface of the eye is thus divided into *facets*, which may be circular if they are few in number, but which are usually crowded together, and then each facet is hexagonal, like the cells of a honeycomb (plate 39).

The number of facets varies through a surprisingly wide range. When the eyes are little used, as in the dark-frequenting insects that we have already mentioned, they may be only two facets, or

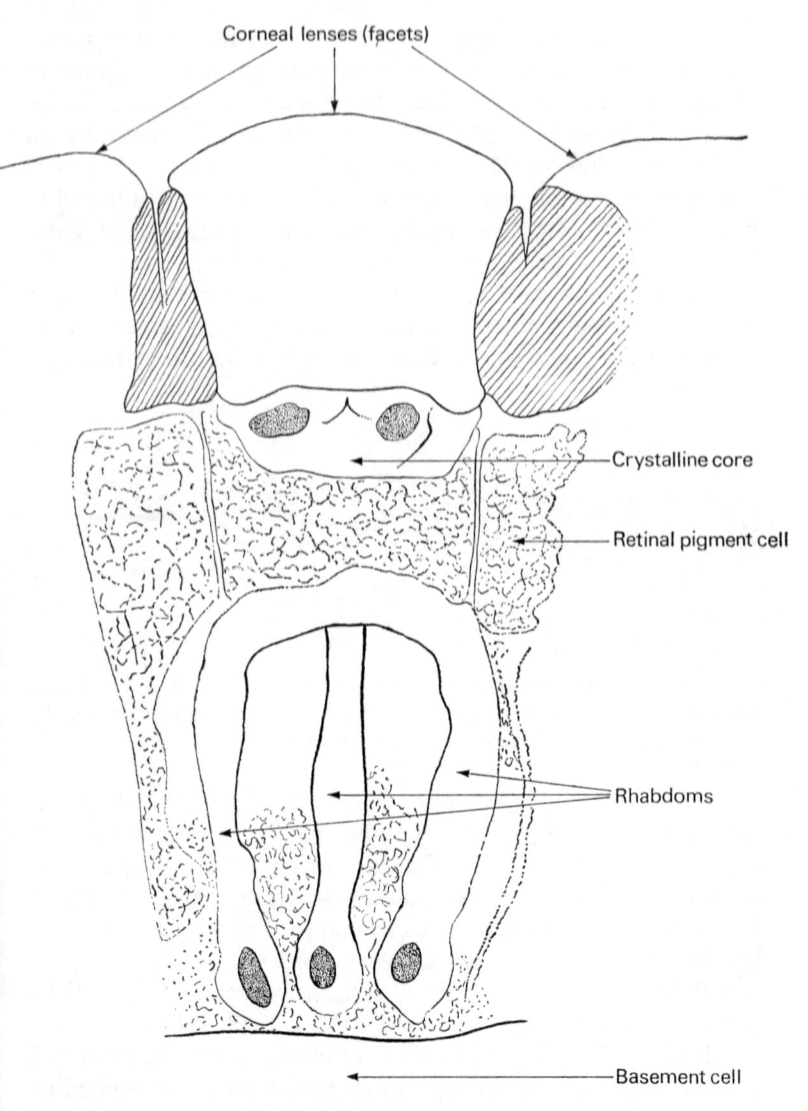

Figure 44. An ommatidium in the eye of a crane-fly (after Sotavalta).

even one. Such very much reduced eyes can be seen in the Nycteribiidae, a family of flies that are external parasites of bats. At the other extreme, some Dragon-flies are said to have nearly 30,000 facets in each eye. In general, those insects that make great use of the eyes tend to have more facets, but this correlation is not always obvious. We can see that Dragon-flies should have many facets, because they hunt in flight, but it is not clear why, if they need 10,000–30,000 facets, and butterflies and moths are estimated to have 12,000–17,000, the true flies should be content with a maximum of about 4,000.

The area of the eyes is often enlarged until the eyes almost or quite meet, usually above, sometimes below the antennae. This is most commonly an attribute of males, and the facets may be larger

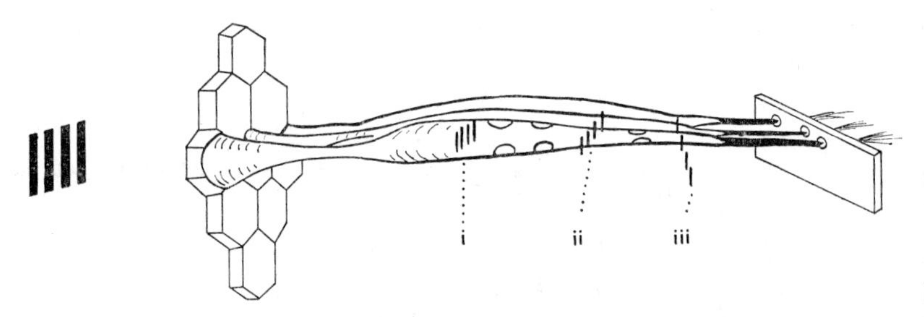

Figure 45. Formation of diffraction images in a compound eye: O, object (striped pattern); E, hexagonal facets of eye; i, ii, iii, successive images formed by diffraction (after Burtt & Catton).

and fewer in the adjacent areas, though the significance of this is not known. In some insects the upper and lower parts of the eyes are physically separated, so that there may appear to be two pairs of compound eyes. This condition has arisen independently in Ephemeroptera (*Chloeus*), in Coleoptera (Gyrinidae and Cerambycidae) and Diptera (*Bibio* and *Simulium*).

A facet is a part of the *cornea*, a transparent area of the cuticle. Beneath this may be a *crystalline cone*, a hard, refractive body, which in combination with the facet of the cornea forms a real image of whatever lies in front of it. Eyes with a true crystalline cone are called *eucone* eyes. Three other types of eye are recognised: *exocone* eyes, with a cone of different origin and nature; *pseudocone*, with a fluid, in place of the crystalline cone; and *acone*, without the cone.

The familiar lenses of glass that we use in cameras, microscopes and the like are made from a material that has a uniform refractive index and the bending of the light rays to form an image is then determined by the curvature of the various surfaces and the distances between them. The cornea and the crystalline cone of an insect's eye are of a laminated structure, made up from layers like an onion, and the refractive index is greatest along the axis and least towards the sides. As a device for bringing light rays to a focus the lens-cylinders of the insect eye are thus more complicated than the simple lenses made by man. Notwithstanding its imperfections, the image formed by a lens-cylinder is a 'real' one in the optical sense, that is rays of light pass through it, so that the image can be received on a surface and can be photographed. A photograph of Professor Poulton, taken through the eye of a butterfly, using a single facet as a lens, by Dr Eltringham has been reproduced in many English books.

Yet it seems that the insect has no means of seeing this image as a picture and merely records it as a spot of light, averaging the light-intensity over the small area that is before that one ommatidium. Each ommatidium has a field of view of six–eight degrees, and overlaps with those of its neighbours in such a way that its centre is about on the edge of the fields of view of the six facets that touch its sides. Any small area of the object is therefore seen by a number of adjacent ommatidia, which produce spots of light varying in intensity according to the light and dark areas of the object. The organ which detects this spot of light is a bundle of nerve cells called *retinulae*, usually seven in number, grouped round a central rod called a *rhabdom* (figure 44). The light falls on the end of this bundle, and eyes can be classified into two kinds according to the arrangement of the retinulae in relation to the lens-cylinder.

In the first group the rhabdom and retinulae are placed in contact with the end of the lens-cylinder, and each lens-cylinder is isolated from the others by a sheath of black pigment. In this arrangement each set of retinulae receives light only from one facet, and the brain of the insect receives from the various nerve-fibres a mosaic of impressions of what is before each facet. This is called an *apposition* eye, because the spots fit together like the spots in a half-tone block in a book or newspaper to form a picture by apposition.

In the second group, the light from each facet is not directed exclusively into one set of retinulae, but on the contrary is encouraged to spread into the adjoining ommatidia. There are various modifications in detail. The pigment layers are reduced; the focus of the lens-cylinder may be shorter so that its image is formed not at the tip of the crystalline cone, but inside it, and the emerging rays are already diverging; and the rhabdom may be longer, with the retinulae not close to the tip of the lens-cylinder (figure 44). Such an eye is called a *superposition* eye, because, although it still supplies the brain with a pattern of spots of light, each spot is formed by adding together, or superposing, light from several facets.

Now the total amount of light that reaches the eye is the same in both cases, but the superposition eye uses it all, whereas the apposition eye absorbs much of it into the sheaths of pigment. The apposition eye gives a clearer picture of the minute changes of density over the surface of the object, and is the better as long as there is enough light. This is the eye for bright light. The superposition eye is better when the light is dim, but because of the spread of light into adjacent ommatidia it tends to blur outlines and fine detail. In many insects that fly at night or at dusk, especially moths, the pigment sheaths may expand and contract and so change the functioning of the eye between apposition and superposition types, by varying the amount that the light strays into adjoining ommatidia. Sometimes this seems to be in direct response to the intensity of the light, but sometimes it seems to be a regular rhythm, or to be part of the general state of rest or alertness: for example, if a night-flying moth is resting by day and is suddenly disturbed its eyes may automatically change to night adaptation, as if it were 'waking up' at night.

It has long been a puzzle why the retinula and rhabdom need to be so long, but it has recently been shown, by very delicate work with electrodes placed in the optic nerves, that other *diffraction images* occur at intervals down the length of the rhabdom, and that these are used to detect small movements in the insect's field of view. This work has also offered an explanation of why bees, for instance, find it easier to detect some kinds of pattern more easily than others.

Powers of the compound eye

It is clear that the eye of an insect is able to detect movement, and to some extent patterns of light and dark areas, but this equipment would not seem to give an insect a clear, sharply defined image of the external world. Of course we do not know that what we ourselves see is real, any more than the insect does. The signals conveyed to the brain produce in ourselves certain instinctive reflex actions – such as closing our eyes, or turning away. By more complicated mental processes other visual impressions also influence our behaviour, both at the present time and in the future, by visual memory. Insects react as if their visual impression also influenced their behaviour, and as if they could remember things that they had seen previously: for example, bees returning to a nest, and using visual landmarks.

We assume that we have a better brain and more efficient eyes than insects have, and that therefore our view of the outside world is more complete, nearer to reality, than is theirs. But what is reality? We do not know.

When we compare insects with ourselves we at once find a way in which our vision is incomplete. Insects see about the same amount of the red end of the spectrum as we do, but they see much further into the ultraviolet. Insects that avoid the light – e.g., ants carrying their pupae into shelter – when offered a spectrum of colours that includes infrared as well as ultraviolet, will avoid all our visible spectrum and a space beyond it, choosing as 'dark' places the infrared and the extreme ultraviolet. In precise terms, their general range is from 2,500 Å to 7,000 Å, compared with the human range of 3,800 Å to 7,600 Å (1 Ångström unit = 0.1 uu (millimicrons) = one ten-millionth of a millimetre).

A number of workers have studied the reactions of insects to colours, one of the latest being Pospisil [84], who worked with the House-fly, *Musca domestica*. Experiments have been broadly of two kinds: those designed to discover which colours an insect can distinguish from each other, and which colours it confuses; and quantitative experiments designed to show the *preference* for one colour rather than another. Experiments involving the recognition of colours and shapes are complex and will be described to some extent in a later chapter on insect behaviour.

There is general agreement that insects react differently to coloured light and to a coloured surface or background. In part, at least, this is a consequence of the way in which such experiments are arranged. Coloured lights are generally compared for their stimulating effect.

They lure a hungry insect in search of food. Coloured backgrounds on the other hand are used in two ways. In some experiments dishes containing food, usually honey or sugar-water, are placed on surfaces of different colours and the insects, usually bees, are trained to associate a particular colour with food. This shows at least that they can recognise the colour, but not whether they prefer it, since they are trained to choose it. The experiments described by Pospisil allowed the flies to choose one colour in preference to another. This was a choice of a surface on which to rest, that is, a place of inactivity.

Hence we might expect a difference in choice between lights, which stimulate, and surfaces, which have a sedative effect. And indeed we do. The House-fly chooses a violet or blue light, but a red background. The preference for violet or blue light is about the same as the preference for white light: i.e., the fly is choosing the colours that seem brightest to it. Similarly, a red background is about as attractive as a black one, which of course gives the minimum of visual stimulation, so in choosing a colour on which to settle the fly is picking the one that seems darkest to it.

Dorsal ocelli

The dorsal ocelli of adult insects are mysterious organs. They have the general structure of an eye of some kind. The cuticle is thickened into a *lens* which is strongly convex, much more so than the individual facets of a compound eye.

Each ocellus has only one lens, beneath which is a large bundle of sensory cells. These are grouped in small bundles, each surrounding a transparent *rhabdom*, and the ends of the cells, or *retinulae* point towards the lens. This structure is essentially the same as that of a compound eye, except that in the compound eye each rhabdom and its bundle of retinulae forms a separate unit, or ommatidium, with its own facet.

The nerves from the ocelli go to the same ganglion of the brain – the first, or protocerebrum – as the compound eyes, but each ocellus has a separate connection. Indeed the single median ocellus has a double link, suggesting that it was originally paired like the posterior ocelli. It seems, therefore, that whatever visual information the ocelli give to the brain is passed direct, and not as part of the general pattern of the compound eyes.

Ocelli resemble the eyes of mammals in so far as the image produced by one lens could be analysed by a large number of retinal cells, but an ocellus is obviously very inefficient in this way. The lens is excessively thick and convex, and cannot be focused, so that it can only give a crude image of objects very close to it. From the behaviour of insects it does not seem that they use the dorsal ocelli for looking at very near objects: for one thing these structures, as their name implies, are placed on the top of the head (cf. the lateral ocelli of larvae, below).

Experiments which have been made in blacking out or destroying one or more of the ocelli are rather inconclusive, but they give general support to the idea that ocelli are used to distinguish light from darkness. They may also be used to measure the intensity of the light, in the manner of a photographic light-meter.

No one really knows what is the evolutionary origin of ocelli, whether they are an ancient kind of eye, later supplanted by the compound eyes, or whether they are organs specially developed for some other purpose. They seem to be a primitive feature, part of the ancestral equipment of insects of all orders, retained by most insects living today but abandoned by many, scattered at random throughout the orders. They are perhaps like the human tonsils, fulfilling a function that was once vital (or else they would never have evolved), but now no longer so.

Among possible uses of dorsal ocelli is that of organs of stimulation, or 'toning-up'. When the light conditions are right for the insect to become active – or, alternatively, dangerous enough to require unusual alertness – the ocelli may send a general signal through the entire nervous system, so that any of the normal stimuli from the outside world will provoke an immediate response.

Lateral ocelli, or stemmata

Although often superficially similar, these differ from the dorsal ocelli of adults in several ways. They are the main, and generally the only, photo-receptors of larvae of holometabolous insects; their nerves lead back to the optic lobes proper; and they are apparently used as the functional eyes.

Stemmata vary enormously in structure from something very like a dorsal ocellus to something like a single ommatidium of a compound eye. It is evident that all three, lateral ocelli, dorsal ocelli and compound eyes, are developed from the same kind of cuticular organ, just as we have seen that many different uses are made of the basic chordotonal organ of the cuticle. There is a transparent area of epidermis, through which light reaches light-sensitive cells, and from these nerves convey a stimulus to the central nervous system. The extent to which these stimuli are integrated into a 'picture' of what is before the insect shows all stages of complexity both in larval and in adult insects.

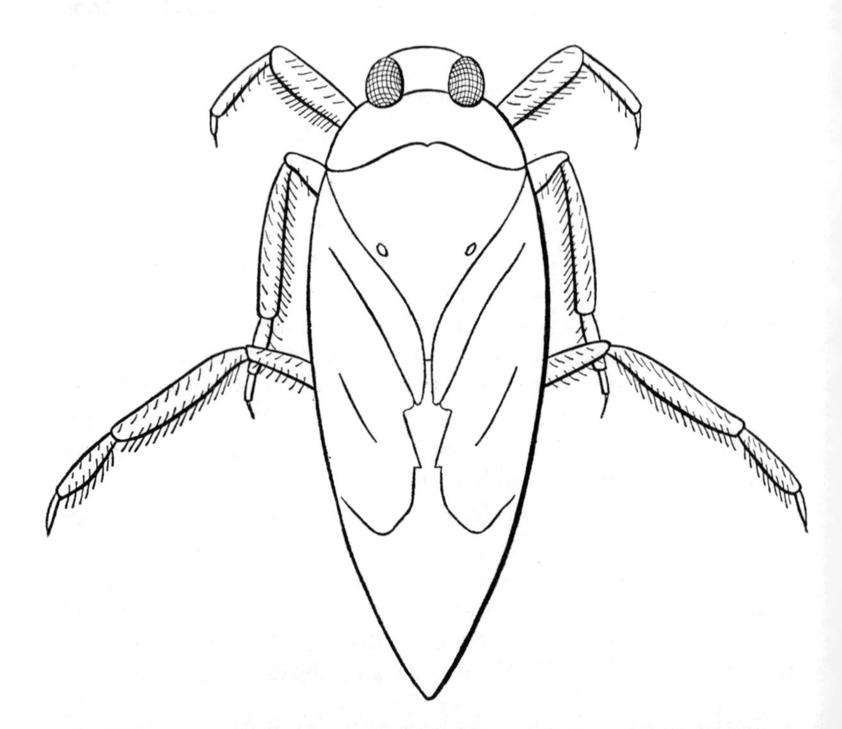

Figure 46. A 'backswimmer', *Notonecta* (order Hemiptera-Heteroptera): an aquatic bug with legs adapted to propulsion in water (after Borror & Delong).

9
Insects and sound*

If it is difficult to visualise the world through the eye of an insect it is even more difficult to decide upon the significance of sound in an insect's life. Insects certainly make sounds; equally certainly some insects can detect sounds. We can point to sense-organs which can respond to the impact of sound waves, and we can sometimes judge from the behaviour of a living insect that it is being influenced by hearing a sound. Yet when we speak of an insect's 'hearing' we must avoid the natural tendency to think in terms of a symphony concert or a political argument. To an insect 'hearing' means very much less than that, and varies from a single impulse of warning or recognition to a fairly simple sequence of notes.

Hearing and the production of sound by insects are only partly associated with each other. Some of the sounds that insects make are incidental noises. Others may frighten or otherwise influence other animals even if the insect itself does not hear its own sounds. On the other hand it may have been an evolutionary advantage to the insect to hear extraneous sounds coming from a variety of sources.

Hearing

Let us start with hearing, because without this the value of sounds to insects is greatly reduced. There is no universal organ of hearing in the insect world. Insects – indeed arthropods in general – modify a simple, basic structure in a multitude of ways for different purposes. Hearing organs have been developed from a relatively

* For much of the information used in this chapter I am indebted to the comprehensive book *Insect Sounds* by P. T. Haskell, to which I make grateful acknowledgement.

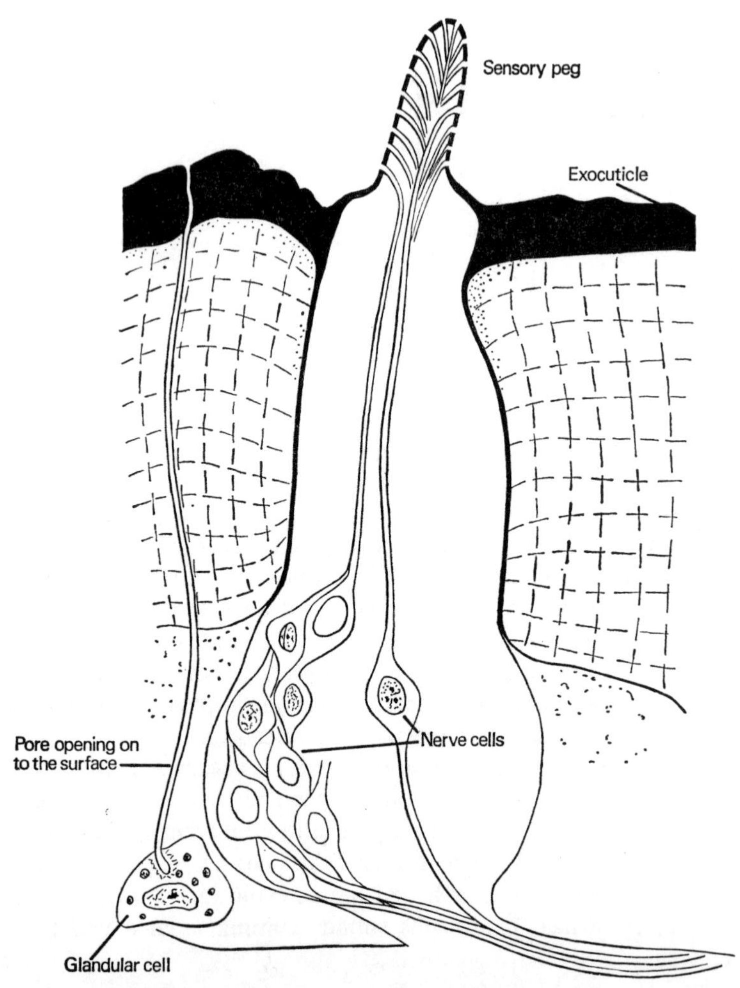

Figure 47. A 'sensory peg' in the cuticle of bees, beetles and moths is connected with the senses of smell and taste (after Schneider).

simple arrangement of nerve cells called a *chordotonal sensilla* (figures 47 & 49). This is closely related to an organ of touch and is attached at one end to a part of the outer body-wall (cuticle) of the insect. Usually the sensilla is stretched between two such points, but sometimes one end remains free. A sound wave arrives as a series of pulses which are transmitted by the cuticle to the chordotonal organs and then converted into a nervous impulse.

Although there are no ears in the ordinary sense, certain special groups of chordotonal organs sometimes occur, and the hearing organs of insects can be arranged in four groups of increasing complexity:—

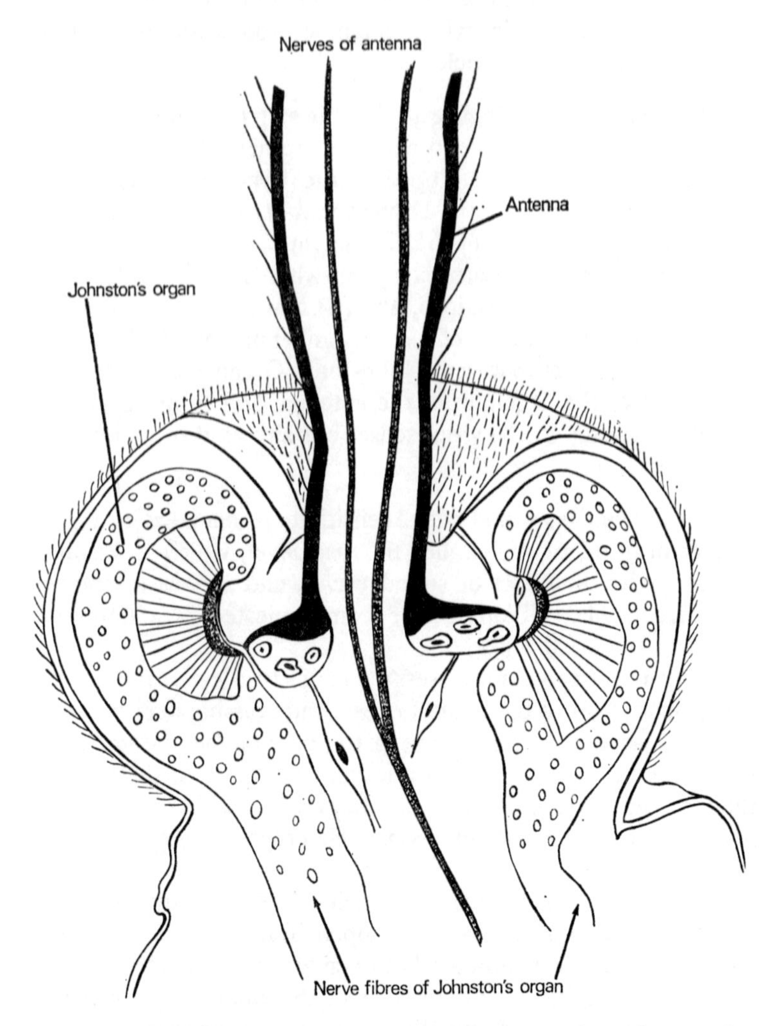

Figure 48. Johnston's Organ in the base of the antenna of a male mosquito.

(a) *Scattered chordotonal sensillae.* These are the simplest organs of hearing, and may occur anywhere, reacting to vibrations of the cuticle.

(b) *Hair sensillae.* Each one of these is a slightly more sensitive device, where a flexible hair growing out of the cuticle picks up vibrations from the air, which might be too weak to affect the general surface of the cuticle.

(c) *Johnston's Organ* (figure 48) is a big step forward in complexity. It is found in the second antennal segment (pedicel) of most insects, and is highly developed in the larger and more recent orders such as Lepidoptera, Diptera and Hymenoptera. It is not different in principle from the simple cuticular organ: its advantages are: (i) that it is located at a point where it is particularly well exposed to external vibrations, and (ii) that it is made extremely sensitive by a combination of a large number of individual sensillae. Haskell points out that what Johnston's Organ really detects is movement of the flagellum of the antenna, and so its function is not only hearing, in the strict sense, but any kind of vibration or air movement.

(d) *Tympanal organs.* Here the sensitivity is increased by having an area of thinner cuticle called the *tympanum*, which moves more readily under the effect of sound waves, and acts as a vibrating membrane. The vibrations are communicated to chordotonal organs of the usual type.

Tympanal organs come closest to being a true 'ear', but they may occur in any part of the body. Locusts and grasshoppers may have a pair of large tympanal organs at the base of the abdomen, and longhorned grasshoppers and crickets have them in the tibiae of the fore-legs. In Geometrid moths they are located in the abdomen; in Noctuid moths and in certain Heteroptera they are in the thorax).

Tympanal organs are closely associated with the tracheal tubes and sometimes with air-sacs developed from these. The result is that the tympanal membrane has open air on both sides, and the sound wave impinges on both sides of the membrane, though by different routes. This means that the balance of pressures from a sound wave on the two sides alters as the insect turns, and so gives it a way of estimating from what direction the sound is coming. A

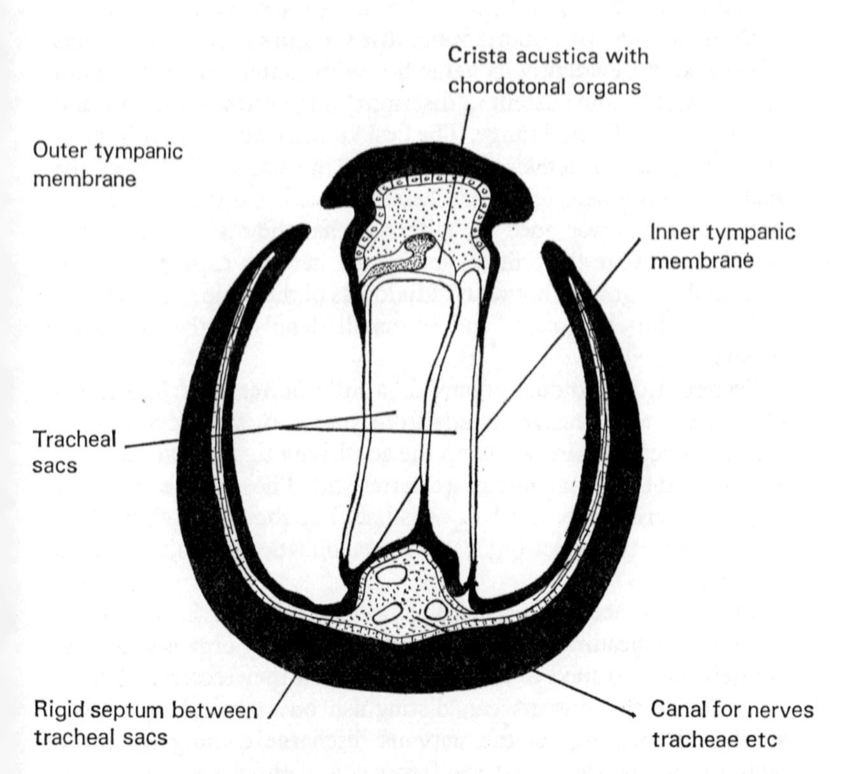

Figure 49. Transverse section across the tibia of *Decticus* (order Orthoptera) in the region of the tympanal organ (after Haskell).

few insects in which this arrangement does not obtain can estimate direction by using the pair of tympanal organs in association with each other.

Range of frequencies heard

Of the four types of hearing organ the narrowest range of frequences is covered by Johnston's Organ: 50–500 cycles per second. This organ must be important because it occurs in so many groups of insects, but evidently its value lies in its being sensitive to faint sounds, and to some extent in discriminating between one note and another over a limited range. The best known example of this use is in the response of female mosquitoes to the sound of flight of the male, but even here recent workers have said that the mosquito is not as clever as was once thought, and that the main effect of the sound is to excite the female, and attract her into close proximity. The final recognition between individuals of the same species is not by sound but by touch, and by small details of the courtship behaviour.

Scattered chordotonal organs do a little better than Johnston's Organ, and may receive sounds from 50–1,000 cycles per second. The presence of hairs sends up the sensitivity to 10,000 cycles per second, and tympanal organs are better still. Those of the knees are about as sensitive as the hair sensillae, but the larger abdominal and thoracic tympanal organs may respond up to 150,000 cycles per second.

These references to precise frequences in cycles per second should not give a misleading impression that the auditory organs of insects are delicately attuned to the frequencies that they receive. There is no evidence that insects can distinguish one musical note from another. Recordings of the nervous discharge coming from the auditory organs show that the frequency of the electrical oscillation in the nerve is determined not by the *frequency* (i.e. pitch) of the sound being received, but by its *amplitude* (i.e. loudness). If a note of steady frequency is varied in loudness, while remaining at the same pitch, the tympanal organs responded to the variations in loudness. They were indifferent to the pitch, provided that this fell within the range that they were able to receive.

Insects therefore are tone-deaf to an extent far beyond that of the most unmusical man. It is useful to bear this in mind when considering the sounds made by the insect themselves.

Sound production

Insects must not be credited with a musical appreciation beyond their powers, and neither must it be assumed that every sound made by an insect is necessarily significant to some other insect. Haskell [41] distinguished three categories of sounds produced by insects:

(a) *By-product sounds*

These arise out of a normal activity of the insect, and so there is no primary reason to assume that they have any 'meaning' at all. It is possible that sometimes by-product sounds are made use of for recognition, or for other purposes.

The best known by-product sounds are those produced by the wings. In the main these are air-movements during flight. Sometimes there are vibrations of the wings when the insect is at rest, and sometimes, as in Hover-flies and in honey bees, a 'humming' or 'singing' tone is emitted by a resting insect, apparently caused by vibrations of the thorax and possibly of the halteres. The wing-frequencies of insects are known to range from 4 to 1,000 cycles per second, and there will be a certain number of higher frequencies arising as harmonics or overtones. In mosquitoes such notes seem to serve some purpose in bringing the sexes together, supplemented at close range by other recognition signs.

Large grasshoppers and locusts in flight produce a note of very low frequency. It is a curiously dry, crackling sound, which has been aptly described as being like the sound of fire in stubble. This sound is said to arise partly from 'wing-collisions', when the hind-wings hit the hind-legs. Acoustically this sound is very irregular in both pitch and intensity, and so possibly may be the sort of amplitude-modulated sound that another grasshopper could well recognise. Yet even so it seems a clumsy and inefficient way of flying, and must waste a good deal of energy. It is difficult to believe that it can be worth the effort, in an evolutionary sense.

These are sounds loud enough for us to hear, but many insects must make by-product sounds that are beneath our notice. A mass of caterpillars feeding produces a cropping sound that perhaps they can hear. A recent triumph of high-frequency recording was to amplify the sound of the feeding of larva of a Hessian-fly in a stem of wheat. The investigators found that the larva 'made its head into a cup-shaped hollow and applied it intermittently to the wheat-stem, while making a sucking noise': surely on a lilliputian scale.

(b) *Impact sounds*

These are the various thumps and scratches that are made by the impact of the hard cuticle of an insect against its surroundings. The best known such sound is the 'ticking' of the Death-Watch Beetle, *Xestobium rufovillosum*, usually heard during the spring months when the mature adult beetles in burrows in the wood are hunting for mates. The sound is the impact of the head against the floor of the burrow, as the beetle jerks itself forward. The smaller 'woodworm', *Anobium punctatum*, makes a similar but less penetrating noise. Certain termites similarly make a drumming noise with their heads, and insects of several orders strike the surface on which they are standing with the tip of the abdomen. Some Orthoptera and Plecoptera do this, and certain Psocoptera of the genera *Clothilla* and *Lepinotus*, though tiny insects, make an audible sound. Pearman has described in detail how the female *Clothilla* (the male does not tap) uses the pull of the legs to raise the tip of the abdomen and then lets it spring back so that the pregenital segment strikes the surface below the insect. He says that the sound is only heard when the insect is standing on a thin and resonant surface such as paper. Many other Psocids have organs on the hind coxae that may be thought to produce sound. Some Orthoptera tap with the tarsi of their legs, and some moths (e.g. *Drepana*) rub their anal segments against a leaf.

It is debatable how far these impact sounds have 'purpose', that is, to what extent the insects concerned have derived an evolutionary advantage from them. Some, like the supposedly audible scratching of the larval mouthparts of the hornet, *Vespa crabro*, must surely be fortuitous. The tapping of the Death-Watch Beetle, made by both sexes, is said to be a mating call, since it takes place mostly at the mating season of the year. No doubt it does stimulate and encourage other *Xestobia* that may hear it, but searching for each other in a maze of larval tunnels must be as difficult as finding the Third Man in the sewers of Vienna.

Tapping by termites is an activity of soldiers, with their large heads. They are said to tap rhythmically and in unison as a danger signal to the rest of the community, and they seem to have developed what was probably at first an incidental noise into a useful activity. Clicking noises are produced by certain moths, and their significance is doubtful. Collenette described a most interesting observation in the forests of Matto Grosso where fighting birds made clicking noises with their beaks, and an *Ageronia* moth arrived on the scene and at each click rose into the air and clicked in reply.

(c) *Special sound-producing mechanisms*

Many insects have evolved one or more organs that certainly produce sound, and which seem to be specially adapted for this function. One must presume that at one time, if not now, the survival value of such a device was sufficient to select and perfect it. I say 'if not now' because it is possible that some of these devices may have been evolved a long time ago, when competition with other insects may have taken place under conditions different from those of the present day. It has been suggested that the flying swarms of insects that one sees about were originally mating swarms, but that today they remain as a persistent habit although mating may be carried out elsewhere. In the same way, insects may go on having organs for sound-production, and even using them, long after their evolutionary value has diminished.

Stridulation

This term is usually understood by entomologists to refer to the production of sound by rubbing two hard parts together: e.g. the legs of grasshoppers against hard veins of the wings. In his recent book Haskell decides that this is an artificial restriction of a useful term, and proposes to redefine stridulation as 'any sound produced by an insect'. Any special mechanism by which an insect produces sound is to be called a stridulatory mechanism. It is true that we need a single word in place of the clumsy 'production of an audible sound', but it is a pity to take away one that already has a precise meaning. Presumably even now it is not meant that stridulation should include by-produced sounds such as the tapping of the heads of beetles and termites, though to be completely logical it should.

Sound-producing mechanisms in this category are of three main kinds:

(a) *Frictional mechanisms* (stridulatory organs in the usual sense).

There are two parts (figure 50). One, the *file*, has a series of ridges or projections, and is drawn across some hard edge or knob known as the *scraper*. The successive impacts of the projections of the file on the scraper set up a series of sound pulses, the interval between which – and hence the pitch of the sound – obviously depends on the spacing of the ridges, and on the speed of movement of the file.

Devices of this type are best known among Orthoptera, in the families Acridiidae (short-horned grasshoppers), Tettigoniidae (long-horned grasshoppers) and Gryllidae (crickets). They are usually restricted to males. The common field grasshoppers rub one or both hind femora against a hardened vein of the fore-wing (tegmen), and at close quarters they can be watched as they stridulate. It has been claimed that the rhythm can be recognised through binoculars, even when the sound cannot be picked out from extraneous noises. A few other Acridiids rub the fore- and hind-wings of the same side together. In contrast, Tettigoniidae and Gryllidae have a different mechanism, clearly evolved independently. They rub the fore-wings, not against the hind-wings, but against each other.

Crickets (Gryllidae) have apparently an even more primitive equipment. Not only do both sexes stridulate, but the mechanism is symmetrical, each wing having both a file and a scraper (plate 16). Tettigoniidae (e.g. Katydids), showing a curious evolutionary economy, have one file, on the left fore-wing, where the anal vein is ridged, and one scraper on the right fore-wing (figure 22). Among Tettigoniidae, wherever there is a scraper there is often also a clear area known as the mirror, which may act as a stretched membranous, resonating surface.

These devices of Orthoptera are the classic examples of stridulating mechanisms. Most of the large orders of insects have evolved such devices, often more than once in the same order. Odonata, Hemiptera, Psocoptera, Lepidoptera, Diptera, Hymenoptera and Coleoptera all have stridulating members, slightly varied in detail, but all similar in principle to those just described.

Beetles have the greatest range and variety of stridulatory mechanisms, not only as adults, but often as larvae. Larvae of the family Passalidae, living in wood throughout their life, have the third pair of thoracic legs shortened and highly modified in shape (figure 50) to form a stridulatory organ which operates against the coxae of the middle legs. Stridulation is evidently so important to

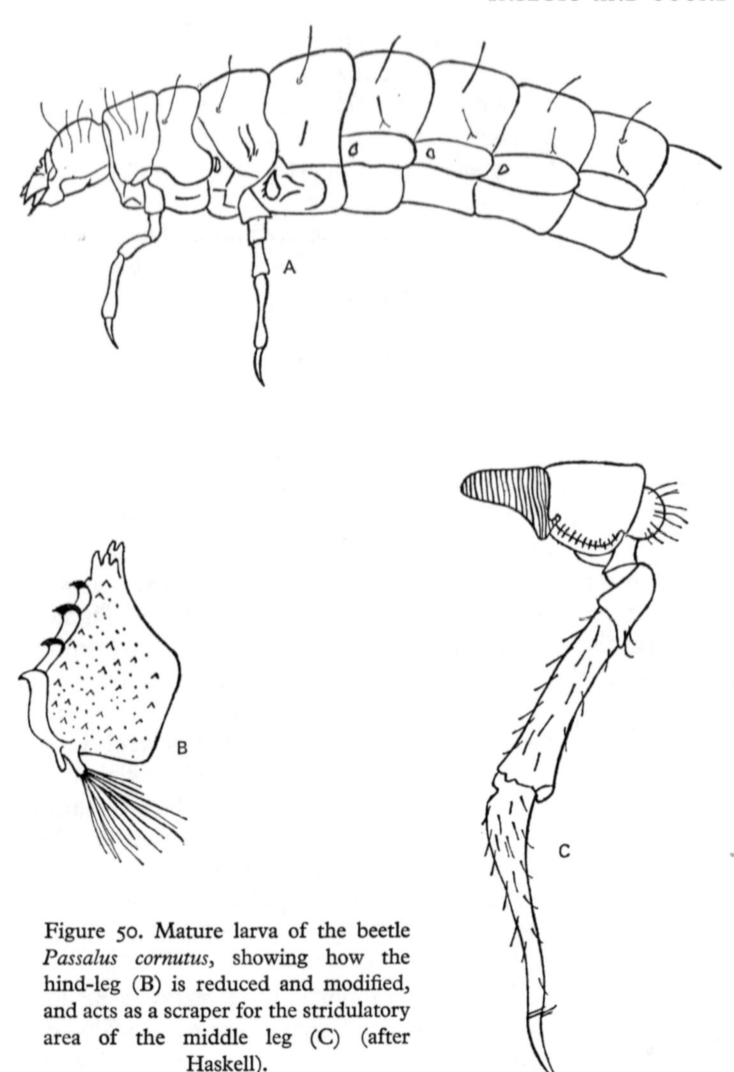

Figure 50. Mature larva of the beetle *Passalus cornutus*, showing how the hind-leg (B) is reduced and modified, and acts as a scraper for the stridulatory area of the middle leg (C) (after Haskell).

these insects that they can sacrifice a pair of walking legs to it. The adult beetles also stridulate, and it is thought that the sound may help to keep the colony together. Even the pupa and pharate adult may stridulate by friction between various surfaces beneath the cocoon, or under the pupal cuticle.

Some Lepidoptera can also stridulate in the pupal stage. Many adult moths rub various parts of the wings, legs or body together, or click a part of the wing-membrane to and fro.

The actual size of these frictional mechanisms is quite small, even in the largest grasshoppers, and it is remarkable what a volume of sound they can emit. This is achieved by transmitting the vibration to as large an area as possible. Crickets and long-horned grasshoppers have special membranous surfaces known as 'mirrors', but the general body surface is a more effective resonator especially if it is tough and relatively rigid. Perhaps that is why beetles, with their heavily armoured cuticle and capsule-like bodies, include some of the noisiest stridulators.

(b) *Vibrating membranes*

We are all familiar with sound-producing equipment that depends on the vibration of a thin membrane, gripped firmly round its margin. The sound-box of the old-fashioned gramophone and the earpiece of the modern telephone use this mechanism.

Certain insects, notably male cicadas, have evolved a pair of similar structures, which they use most efficiently. These are situated at the base of the abdomen, and in each organ a membrane is held by an elastic, sclerotised, supporting ring in such a way that the *tymbal* membrane is permanently bowed outwards. A muscle attached to the membrane contracts and relaxes several hundred times a second (120 to 480, according to Pringle [85]). At each contraction the membrane is flattened, then when the muscle relaxes the elastic ring restores the bulge in the tymbal. Other, ancillary muscles attached to the elastic ring distort its shape, and in this way affect both the volume of sound and its *timbre*, i.e. the admixture of overtones from which the characteristic 'song' is derived.

The loudness of the sound is amplified by resonators: sometimes resonating chambers, or *opercula*, derived from outgrowths of the cuticle of the last thoracic segment; sometimes air-sacs developed in the tracheal system. Even with these aids it is remarkable that such a great volume of sound can be produced by such a small creature as even the biggest of the cicadas, though these are giants among living insects.

Certain moths of the families Arctiidae and Syntomidae have a membrane covering a cavity in the thorax. During flight the thorax is distorted by the flight-muscles, and this produces a loud crackling sound. Such a device is not on the same level of evolutionary ingenuity as the tymbal of the cicadas, since it cannot produce a 'song' independently of the wing-movements. Indeed it should rank as a by-product sound.

(c) *Sounds produced by air-movement*
All the insects we have so far considered have set up pressure waves by very rapid movement over a short distance, and there was no substantial movement of air. Some moths, such as the Death's Head Hawk Moth, produce an audible squeak by puffing and blowing through the proboscis, and certain others are suspected of making a noise by the rush of air through the thoracic spiracles. This has been said of the 'humming' of Hover-flies when at rest, and of the 'piping' of queen bees. In neither case is it certain that the sound is not primarily caused by vibrations of the thorax, with movement of air as a secondary effect.

The uses of hearing and sound-production
No one really knows whether sound is important to any particular insect. The ability to hear is presumably always an advantage, since it adds to the total of information received from and about the outside world. Yet the comparatively poor hearing, and in particular the lack of discrimination, among insects suggest that hearing does not play a critical rôle in an insect's life. Sound production, in so far as it is purposive in the evolutionary sense – i.e. it is developed and improved by natural selection – may be directed towards others of the same kind, or towards outsiders. Suggestions of a possible useful function include:

(a) *Outside the Species*
(i) *As a weapon of defence*, warning and frightening off other animals. This also must imply the drawback of attracting attention, and inviting attack.
(ii) *As an aggressive challenge* to other males, establishing a territory as birds do. This is believed to be true of the song of male crickets.

(b) *Within the same Species*
(i) *Mutual recognition* is undoubtedly helped by the production of characteristic sounds. Local populations may develop their own

version of the specific song, and this in time adds to the *reproductive* isolation of local populations from which new species arise. Some species must be at such a stage of their evolution at the present time, since local forms can sometimes be recognised by their song when they show no apparent structural differences.

(ii) *Assembling*. The communal song of cicadas or crickets helps to keep the individuals within hearing of their own kind, and to discourage dispersal. No doubt this reduces the likelihood that the population will be weakened by scattering, but it is also deprived of the chance to venture into new and stimulating environments. Perhaps it is significant that the insects best known for their powers of stridulation – Orthoptera, cicadas – have survived for a long time, but have changed little.

10
The flight of insects

Flight is one of the outstanding characteristics of insects as a class, and one that they share only with birds and bats. With the solitary exception of Ephemeroptera (the 'dun' or subimago of the May-fly) it is a characteristic of adult insects, fully mature, and having completed their final moult.

Origin of wings

The origin of wings in insects has been much debated, and is generally thought to have occurred only once in evolution. Lemche [64] believes otherwise, but even he does not suggest that wings have evolved on more than two or three separate occasions in the long history of insects. This is rather remarkable when it is considered how many times insects of widely different orders have arrived at very similar structures by convergence: for example, the many similar devices for life in water (Chapter 12). Yet the wings of all winged insects are clearly built on the same basic plan, in spite of considerable modifications of detail.

The wings are sac-like growths arising from the cuticle at the junction between the notum and the pleura of the thorax (figure 1), with the upper and lower surface pressed together and supported by the branching veins (figure 17). The wings of all insects are elaborately hinged to the thorax, with the result that they can be twisted, and the wing-tip moved round through a closed loop, circle, ellipse, figure-of-eight, or any variation of these. Twisting the wing as it moves round its path varies the angle at which the wing strikes the air – the 'angle of attack' of aerodynamic theory.

In a conventional aeroplane with a fixed wing, in steady flight, the angle of attack remains constant. The flow of air produces an increase of pressure beneath the wing and a decrease of pressure over the upper surface, the two combining to exert a force at right angles to the wing as shown in figure 51. This force can be resolved into two components at right angles: the *lift*, which acts upwards, and opposes the *weight* of the aircraft, and the *drag*, which combines with the other forms of air-resistance to oppose the *thrust*, or force which is impelling the aircraft forward. During steady forward flight, without gaining or losing height, the lift balances the weight, and the thrust balances the drag.

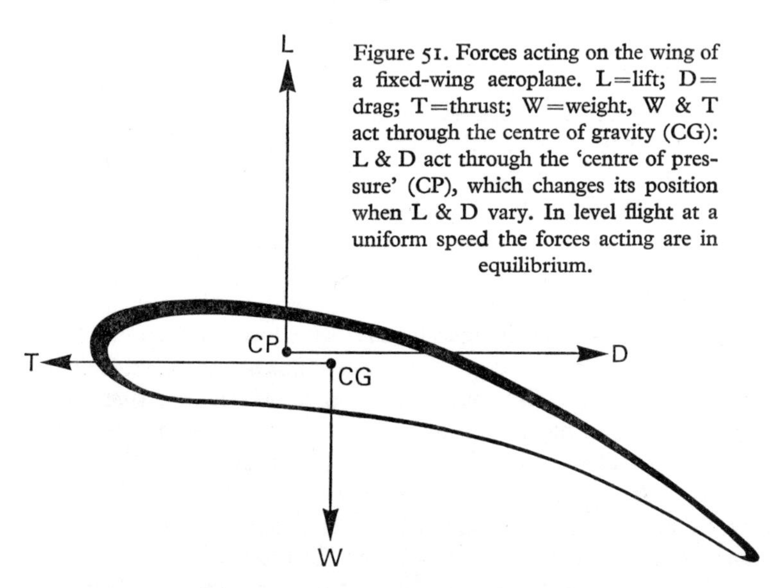

Figure 51. Forces acting on the wing of a fixed-wing aeroplane. L=lift; D=drag; T=thrust; W=weight, W & T act through the centre of gravity (CG): L & D act through the 'centre of pressure' (CP), which changes its position when L & D vary. In level flight at a uniform speed the forces acting are in equilibrium.

The great difference between insects and aeroplanes as flying machines is that in aeroplanes the wings provide only lift, and the forward thrust is supplied independently by propellers, or by jet-motors. In insects the wings fulfil both of these functions, and thus are analogous more to the rotors of a helicopter than to the fixed wing of an aeroplane.

It is fairly certain that the wings of insects first arose as fixed planes, which projected from each side of each segment of the thorax and probably from the abdominal segments as well. At this stage the insect flew in much the same way as a man-made glider, using the lift supplied by the wings to reduce the rate of descent, and so converting a free fall into a long glide. It has been plausibly argued that wings first became capable of turning, and thus altering the angle of attack, making the angle of glide shallower or steeper as required. This type of flight is the same as that of a conventional aeroplane, except for the lack of engine power.

The further step of flapping flight was a big one, especially as all the power had to be applied at the base of a clumsy wing. Much bigger muscles were required than those which sufficed merely to twist the wings, and really positive flight, including the ability to take off from the ground, must have been delayed until the necessary muscular system had been evolved. For a long time insects must have climbed to a height before each flight.

Recently Wigglesworth [122] has disputed this classic picture of the origin of wings in insects, and has suggested instead that wings may have evolved first among the small insects that are carried about by air currents. This method of dispersal has been given much attention in recent years. For example Gressitt [36] has stressed that this is a more likely explanation of the world-distribution of insects than the hypothetical land-bridges and drifting continents that many zoogeographers have invoked.

Wigglesworth points out that even many wingless insects are carried by air current: '... small hairy caterpillars ... one human flea ... wingless Thysanura, Collembola, young stages of Hemiptera and Orthoptera, larvae of Coleoptera, Lepidoptera and Diptera, and great numbers of wingless ants ...' Although for most of the flight such insects drift helplessly, he considers that the possession of wings would help insects to take off and land (though we have just commented that control at these times must have come *after* control in flight), and that this would give them an evolutionary advantage. It would increase the frequency with which that particular group of insects would be carried to new areas and would be able to land safely.

Wigglesworth's objection to the idea that flight arose among large, slow, gliding insects is that he cannot see how in such insects 'very small steps towards the evolution of completely functional wings could at once have selective value'. I must confess that I do not see why this problem should be any different in insects of different sizes. Anything that helps to keep the insect in the air could be turned to evolutionary advantage.

Mechanism of flight

Leaving aside the problem of the origin of wings, the ability to move the wings altered the nature of insect flight from that of a simple glider to something more like a helicopter. To produce lift a wing needs a 'relative wind': i.e. a flow of air over the surface regardless of how this flow comes about. An aeroplane with fixed wings has to be dragged through the air to produce a relative wind, and so can fly only as long as it is moving forward. An insect and a helicopter produce a relative wind by moving the wings, and hence can produce lift and fly without necessarily moving forward.

The wing-tip of a flying insect follows a closed path that is different in still air and in an airstream (i.e. during flight). It is obviously difficult to photograph or otherwise record this path if the insect is moving in free flight, and so most experimental work has been carried out on tethered insects. It is then necessary to provide a current of air to get the same effect as if the insect were moving. The complicated movements of the wing during this cycle have been analysed, partly by direct recording and partly by study of the structure and physiology of the muscular systems involved. A very detailed account is given in Pringle's book [85], and the following is only a brief outline, necessarily oversimplified.

The driving power of the wings comes from the indirect flight muscles, those very large muscles that occupy nearly all the space inside the thorax in many, or most insects. One set, the *dorsoventral muscles*, connect the tergum or notum with the sternum, and when they contract the notum is pulled down and flattened. Working in opposition are the *longitudinal muscles* which run horizontally between the intersegmental partitions; when they contract the tergum is arched upwards. In the simple view, the wing pivots on top of the pleuron as in a rowlock and the movement of the notum operates the wing rather like an oar, though with more up-and-down movement and less to-and-fro action. It will be seen that during the middle part of the stroke the side-walls of the thorax (the *pleura*) must be pushed outwards, and then come in again as the wing moves in to its upward or downward position. Recent research has shown that the degree of rigidity of the pleura has an important effect on the movement of the wing.

In many insects, including many flies, the pleura are stiffened by ridges to which are attached powerful *pleurosternal muscles*. Furthermore, the hinge of the wing is 'double-jointed', being partly

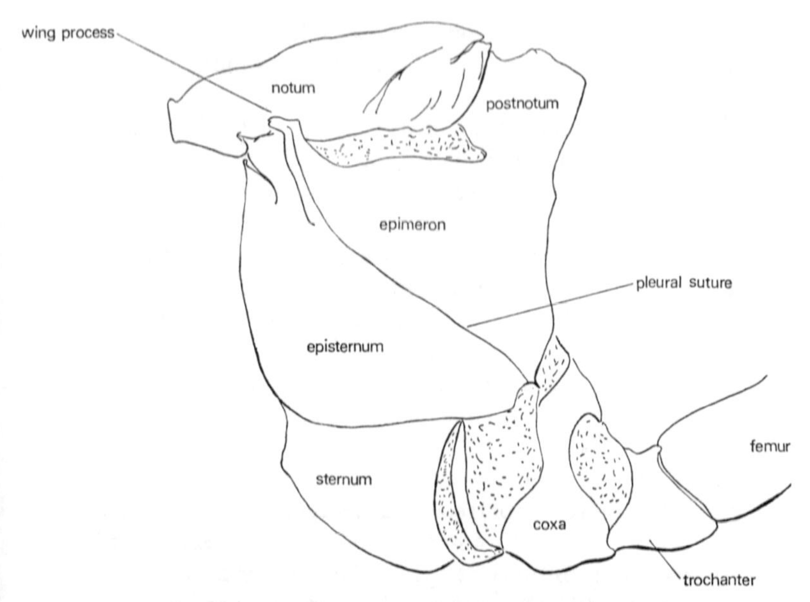

Figure 52. Side view of the thorax of a stone-fly (order Plecoptera), showing the wing-process on which the wing is articulated (after Snodgrass).

supported by an arm or lever which reaches forward from the scutellum, and which moves up and down as the thorax is distorted by the flight muscles. The detailed mechanism is too complicated to describe here, but it is explained very lucidly by Pringle [85]. The effect on wing-movement is to introduce what is called a 'click mechanism': the wing does not move up and down smoothly, but clicks from one position to the other. A similar mechanism can be seen in the ordinary tumbler switch used for the electric light, which clicks from one position to the other to avoid excessive sparking at the contacts.

The complex of articulated *axillary sclerites* and their pivots on the thorax, ensure that as the wing moves up and down it automatically turns on its axis, pronating and supinating like a human fore-arm. This is equivalent to feathering an oar on the return stroke. The most complex articulation is found in advanced insects such as flies, where all the work is done by one pair of wings. In more primitive insects such as Orthoptera or Odonata, where the two pairs of wings still operate independently, there is a phase difference between the two pairs. The hind-wings finish their downstroke while the fore-wings are still in the middle of theirs. It is said that this ensures that the hind-wings move in air as yet undisturbed by the fore-wings.

According to Pringle the increased complexity of the mechanism in advanced insects, and in particular the click mechanism, is correlated with the increased speed of wing movement. As long as wings flapped slowly they could be driven by ordinary muscle responding to a nervous impulse at each contraction. When insects became smaller they needed higher and higher wing-frequencies to generate enough power for flight, and it became impossible for ordinary muscles to contract and recover quickly enough. A new type of muscle was evolved which contracts suddenly when it is suddenly stretched, and relaxes equally suddenly if the tension is released.

This is the function of the click mechanism of the wing. The longitudinal muscles contract and depress the wing, but as the wing clicks over its mid-point the longitudinal muscles are suddenly released from tension, while the dorso-ventral muscles are given a quick pull. This causes the former to relax, and the latter to contract, pulling the wing up again. As it clicks over the opposite effects are produced.

This means that the main power plant of the wings runs automatically once it has been started. An impulse from the central nervous system activates the indirect flight muscles, putting them into a state of excitation, and the alternate contractions of the two sets of muscles then continue until the exciting impulse stops. The frequency at which these muscles contract, and hence the frequency of the wing beat is not controlled by the insect, but is dictated by the size and shape of the insect, and by certain physical factors, particularly temperature.

What the insect does control is the path of the wing-tip, and the angle of attack of the wing at different points on its path; This is the function of the direct wing-muscles, which run from attachments on the pleura to various axillary sclerites. All the elaborate manoeuvring of a Hover-fly or a bee is achieved by pulling on these tiny sclerites, often quite differently in the two wings. As in a helicopter, the resultant effect is to produce a reduction of air-pressure above the insect, thus supporting its weight, and to move the partial vacuum ahead, astern, or to one side at will. The insect is pushed bodily towards the reduced pressure, and this is how a hovering insect can remain in one place and then suddenly displace itself to any direction at a sign of danger.

This sounds dull when so expressed, but remember that the insect receives a brief sensory impulse – a shadow across the eye, or the feel of a tiny current of air – and translates this instantly into outgoing motor impulses which move the axillary sclerites, and flex the two wings independently in such a way that the insect is carried away from danger. Of course we do something similar when we assess the flight of a tennis ball and move our own racket to meet it, but in comparison with the flight-movements of an insect we are performing in extreme slow motion, and even then we do not always succeed.

Most insects take to flight if they are alarmed and so any sudden movement that they can see, or feel will put many insects to flight (some, however, do the opposite and cower down, making themselves as invisible as possible). Nearly all insects start to fly if their legs lose contact with a solid surface beneath them, and in some experiments it was found that legless insects could not stop flying. Insects behave differently when suspended in the air without contact for their legs. Some, like the Fruit-fly, *Drosophila*, may continue to fly until exhausted, but others soon stop unless they feel a flow of air over the antennae, as they would experience it during flight.

11
Adaptation to terrestrial environments

General

An animal, like a machine, has a certain basic structure, which confers upon it advantages and disadvantages. If a tool is made for one purpose it automatically becomes unsuitable for certain other purposes, and the more perfectly it is designed for one job, the less easily it can be adapted for other uses.

This problem of *adaptation* is a fundamental one in the evolution of animals. All through their lives animals are in competition with others, including other members of their own species. A Blow-fly may lay a thousand eggs during its lifetime, and unless the total number of Blow-flies is to increase catastrophically, all but two of these must perish before they reach maturity. The process of elimination leading to these two survivors is the natural selection that is the core of Darwinian teachings on evolution. Though pure chance plays a big part, there are also many times when tiny individual differences in structure or in habit make the difference between those that survive and those that perish. Thus Wigglesworth, in his latest theory of the origin of insect flight, suggests that tiny variations in efficiency of the wing-moving mechanisms among small insects that are carried about by the wind may have finally resulted in the evolution of fully controlled flight.

All animals are therefore subjected throughout life to a selective pressure which tends to mould the structure of the animal, and also its behaviour, to fit the needs of the immediate environment. This would be wholly beneficial if the environment remained eternally the same, when animals would steadily become more and more perfectly adapted to it. The environment itself changes continually, and animals that have committed themselves too far in one direction – have become too *specialised* – may find that they cannot remodel themselves to suit the new conditions. They decline and perish, to be replaced by some other group which have remained *generalised*, less efficient perhaps than the others, but with more capacity for change.

This process, repeating itself endlessly, is evolution. A hitherto generalised group adapts itself to suit a particular way of life, flourishes in its day, and then is left behind by changing conditions, while others take its place. This is the story of the adaptation of insects to varied environments, to living in the soil, in wood, in water, on or inside plants of various kinds, to exploiting all the so-called 'ecologicial niches' in which it is possible for insects to live. It is also an important element in the distribution of insects about the earth, a process that is going on continuously even at the present day. Though insects may be carried passively across deserts and oceans by wind, by water, or by man's agency, they can only survive in their new country if they are able to adapt themselves to its conditions, and its fauna and flora.

An insect is a cold-blooded, air-breathing animal. Even the biggest insects of the present day are really small creatures, with a relatively large body-surface which aggravates the effect of unfavourable surroundings: the loss of heat in very cold places; the loss of moisture in arid regions. The peculiar properties of the insect cuticle are partly a protection against these two dangers.

The tropical forest

Insects live in such a wide range of environments that it is difficult to say what is the optimum, the ideal place for an insect to be in, but probably it could be said that the humid tropics present insects with the fewest problems. There conditions remain about the same throughout the year. There are no great variations in the length of day and night, or in the temperature and humidity during the twenty-four hours, or even during the year. In recent years entomologists interested in the behaviour of biting insects, particularly mosquitoes, have studied them at all hours of night and day, through all the seasons of the year, and – by means of high steel towers – at all levels up to the highest canopy of the forest trees. Even in the tropical forest variations in temperature and humidity do occur, but an insect can always avoid these if it does not like them, by moving up or down, by taking shelter, or by coming out into the open. For insects as for man, the tropical forest offers an easy way of life up to a point, but it easily leads to stagnation. Outside, on the forest fringe, and in more open grasslands and woodlands, there has been more stimulus to adaptation, and hence to evolutionary progress.

Conditions resembling those of the humid forest of today existed for a long time indeed, perhaps for 60 million years, during the Carboniferous period, and among fossils of this period are to be found the first examples of winged insects. Not only was there a vast area of tropical forest, but its distribution was quite different from the tropics of today. Coal measures are found in Spitzbergen, and in the Arctic, showing that for a long time, long enough for these great forests to be repeatedly submerged and fossilised, these countries had a climate at least warm enough for such forests to flourish.

Microclimates

When talking about climatic conditions as they affect insects, one must remember that each insect is concerned only with its own immediate surroundings, its *microclimate*. If a small local patch, a hollow beneath a stone, or a hole in the ground, has a suitable temperature and humidity, always remaining within tolerable extremes, an insect will live there contentedly, without caring whether it is surrounded by tropical forest or sandy desert. Thus even in a tropical forest many insects live in clearings, along the edges of streams or rivers where the light can penetrate, or up in the high canopy, where they have light and air and free movement, as if the dark forest below did not exist. Similarly, true forest insects can find themselves microclimates of a forest type far outside the boundaries of the continuous forest area: for example in the 'gallery forest' along the beds of streams, and in such out-of-the-way places as the little patch of rain forest that has formed in the spray from Victoria Falls in Rhodesia.

It must be remembered, too, that for purposes of discussion we divide past history into 'periods' which are just as artificial as the 'periods' of the history that is taught in schools. Many English children leave school with the curious idea that in the year 1603 one lot of people called 'The Tudors' all died and were replaced by another lot of people called 'The Stuarts'. Of course this was not so. Many people, and certainly many families, lived in both periods.

Yet the times were different, and favoured different qualities in people, so that there was a social evolution from one period to the next. In the same way, as families of insects lived on from Carboniferous to Permian times there was no sudden change – except on a geological time-scale – but there was a gradual shift in emphasis in environmental factors, which favoured some groups of insects at the expense of others.

Although the climate of the Carboniferous period may have favoured large, slow-flying insects, there must have been areas where it was cooler and drier, and where smaller insects with more efficient flying mechanisms would be at an advantage. The disappearance of the huge Carboniferous forests in succeeding epochs gave these groups a stimulus to expand and diversify, while the ancient types that were too heavily committed to forest conditions declined. At the present time some very large insects of the tropical forests – the Goliath Beetles of Africa, and the Pantophthalmid flies of the Amazon, for example – are apparently dependent in this way upon the continuance of their environment, and would be unable to adapt themselves if the tropical forests were to disappear.

Some of the insects of the ancient tropical forests did succeed in adapting themselves to change. Cockroaches have survived almost unchanged since Carboniferous times, and the May-flies (Ephemeroptera) and Dragon-flies (Odonata) of the present day must have arisen from ancestors of that period.

Adaptation to other conditions

The need to adapt themselves to more varied and more rigorous climates affected both the adult insects and their immature stages. The perfection of flight was one reply to this stimulus. Whether it arose as a gradual improvement of the gliding powers of insects of the Carboniferous forests, or among small, wind-borne insects, as Wigglesworth suggests, the power of flight certainly helped insects to move further afield in search of suitable places in which to live, and helped them to cross unfavourable tracts of country on the way.

Adaptation of mouthparts

Early insects had mandibulate mouthparts, with which they chewed either plant tissues or other insects. Insects of some orders have always kept this type of equipment. Dragon-flies, even today, are fiercely rapacious both as nymphs and as adults (figure 62). Mayflies, the other ancient group, have ceased to feed at all as adults, though their nymphs still chew underwater vegetation.

Grasshoppers and cockroaches and their relations (plates 13–19) have kept their primitive chewing habits. With their considerable powers of running and jumping, as well as of flying, they were well-equipped to survive among the varied conditions that followed on the break-up of the Carboniferous forests and, they seem early to have found a successful formula and persisted in it. Since grasshoppers of one species or another are at home in any conditions from lush grass meadows to stony and sandy deserts they have always found somewhere to their liking.

Another outstanding example of the retention of the primitive feeding habits is that of the beetles, which have lived in much the same way for nearly 200 million years. From the Archicoleoptera of the Upper Permian to the ubiquitous beetles of the present day, they have kept their ground-living habits and their chewing mouthparts.

Outside these orders, chewing mouthparts are retained by some rather obscure groups like the chewing lice (Mallophaga), and by the *social insects:* at one end of the evolutionary scale by the termites and at the other by the bees, wasps and ants. They make great use of their mandibles in nest-building, in defence and in passing food to-and-fro (trophallaxis), and it may be that the practical advantage in this respect has caused mandibles to be retained even among bees, which have also developed a long sucking proboscis from the glossae of the labium (plate 27).

Most of the more advanced insects, however, have modified their chewing mouthparts into some kind of sucking proboscis. This has occurred in the bugs (Hemiptera), the most successful order of Hemimetabola, and again among Holometabola in butterflies and moths (Lepidoptera) and flies (Diptera). Smaller orders of sucking insects include sucking lice (Siphunculata) and fleas (Siphonaptera). They do not all use the same modifications. Flies make a tube from the mandibles and maxillae, like the very distantly related Hemiptera, whereas the much more nearly related Lepidoptera suck through the galeae, and bees through the glossae (plates 27, 31, 39).

It is clear from this that a sucking proboscis must have been evolved independently among many different groups of insects.

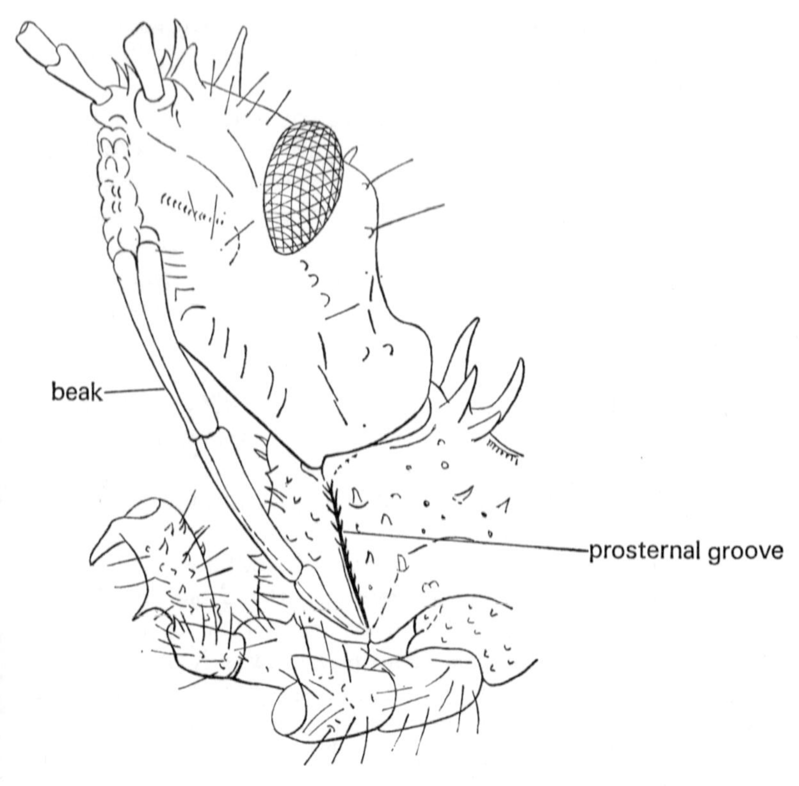

Figure 53. A piercing and sucking proboscis developed by a reduviid bug (order Hemiptera-Heteroptera) (after Borror & Delong).

starting as far back as the 'Bug-nosed Dragon-fly', *Eugeron*, of the Lower Permian. It is to be expected that mouthparts would be very much subject to adaptive change, since any improvement in their efficiency would immediately benefit the insect concerned. It remains surprising, however, that insects could evolve a sucking proboscis several times, whereas effective wings seem to have arisen only once: all winged insects come from a common ancestor, but this is not true of all sucking insects.

The benefit of a sucking proboscis is that it gives access to more concentrated forms of food. A chewing insect feeding on plant material suffers from the disadvantage of any herbivorous animal, that its food is bulky. It must masticate a great number of plant cells to get enough nourishment, and therefore it needs a capacious intestine, usually with a crop or reservoir. Unless it has assistance from intestinal Protozoa, as do many termites, the cell-walls are only useless roughage.

As soon as an insect has evolved sucking mouthparts it can take liquids without bulky solids. If it can pierce as well, it can extract the sap of plants, and this was the oldest adaptation and the most universally successful. The colonies of aphids encrusting juicy young shoots in early summer are a wonderful advertisement for sap-sucking. There is no need even to move about: the insects can stay in one place, and the plant will maintain a steady supply of sap, at least for a time.

Even this can be improved upon. The body-fluids of other insects, and the blood of vertebrates, are each still more nutritious than sap, since they are rich in dissolved nitrogenous foodstuffs on their way to supply the host animal. Reduviid and Polyctenid bugs have long ago changed from plant-feeding to this more rewarding diet.

Sap-sucking from plants came early in evolution, among the hemimetabolous orders. The more advanced Holometabola also reverted to a sucking habit, but in response to the appearance of the flowering plants of the Cretaceous period. Here is an interdependence, a *symbiosis*, or a 'mutualism'. We take the existence of flowers for granted, and forget that they exist only because they improve the prospects of successful fertilisation by attracting insects to settle on the flower (figure 54). They attract insects only because they provide nectar, a concentrated sugar solution, and pollen, a source of vegetable protein. Thus the Panorpoid orders (see Chapter 6) on the one hand, and the bees on the other, began as mandibulate forms, chewing their food, and under the stimulus provided by the flowering plants they independently evolved at least three different kinds of sucking proboscis: one for butterflies and moths, one for flies, and one for bees. No true flies have retained their chewing mouthparts; nor have any true Lepidoptera, since the mandibulate Micropterygidae have now been removed to a separate order Zeugloptera. Among Hymenoptera, in contrast, even bees have mandibles, and wasps and ants have kept their mandibles and their chewing habits entirely (plates 42, 44).

Once again, as among bugs, some flies have discovered the advantages of sucking the blood of other insects or of vertebrates. Apparently this 'discovery' was made quite early in the evolution

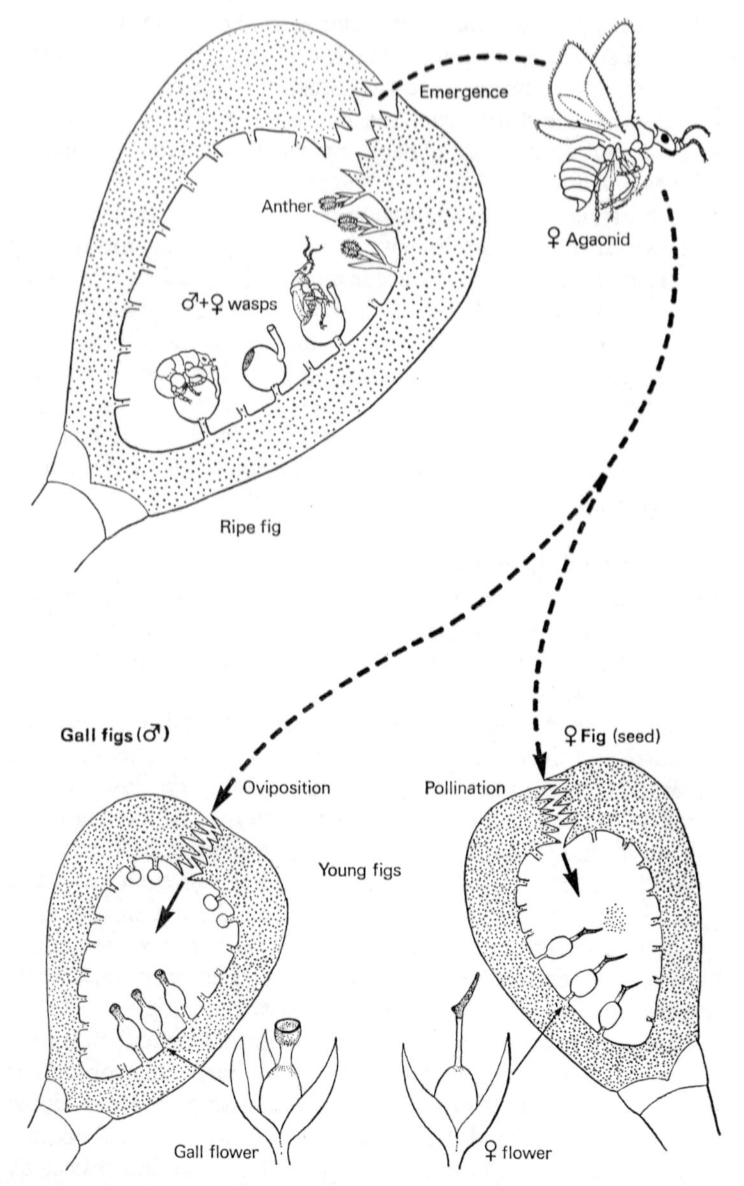

Figure 54. Diagrammatic representation of the life cycle of an agaonid fig-wasp (order Hymenoptera) in a dioecious fig species (i.e. one with male and female flowers separate (after Hill).

of flies. Mosquitoes, Black-flies, biting midges, and Horse-flies form a group of families among which the biting habit is apparently retained from a primitive ancestor. Most other flies have long lost both mandibles and maxillae, and hence the ability to pierce, but they still pursue a sucking habit by using the labium with its *labella* as a sponge.

A still further phase of adaptation in this very adaptable order of insects is that certain families have reverted to piercing and sucking to obtain their food. Having once lost mandibles and maxillae they could not get them back, but had to make do with what was left. The predatory Robber-flies (Asilidae) and Dance-flies (Empididae) use the hypopharynx as a piercing organ, while the formidable Tsetse-flies (*Glossina*), the Stable-flies (Stomoxys), the Louse-flies (Hippoboscidae) and the Bat-flies (Streblidae and Nycteribiidae) all use the hardened stem of the labium to pierce the skin of their victims.

Adaptation of the immature insect

The most important way in which insects have adapted themselves to meet the challenge of their environment, and equipped themselves to live in a great variety of different ways, is by the device of metamorphosis.

In insects, as in other animals, the fully mature adult is the 'real' individual, and the immature stages were originally no more than a temporary condition of being unfinished. Wingless insects have little or no metamorphosis: they hatch from the egg looking like a small adult, and although they may moult a great many times they do not change much in appearance, and may not even grow any bigger. In fact if short of food they may moult to a smaller size. They may also continue to moult at intervals after they are adult. Moulting is apparently a device to get rid of accumulated waste materials, and to renew the worn surface of the cuticle.

The evolution of wings, with their complicated articulation with the thorax, made moulting difficult. Quite early in evolution it seems to have come about that functional wings and functional sexual organs belonged to the last instar, which thus became abruptly different from the earlier instars. The adult insect became clearly distinguished from the immature, or 'young' insect. The biological functions of the adult insect are to reproduce and to deposit eggs as far afield as possible, seeking out all possible places

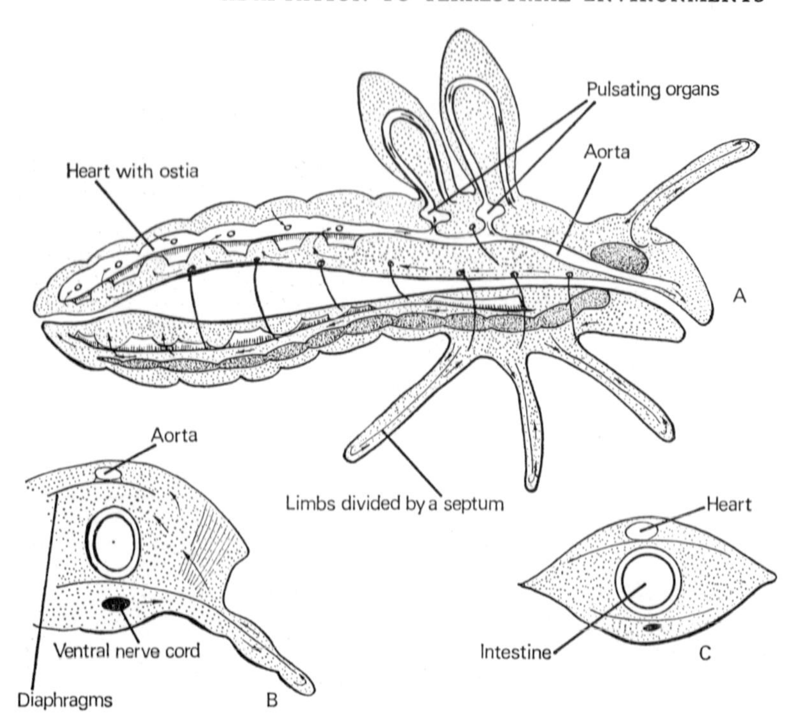

Figure 55. Blood circulation of an insect: A, general view from side; B, cross-section of thorax; C, cross-section of abdomen (after Ross).

in which the insect could survive. The action of the adult insects in spreading into new environments encourages the evolution of new species by natural selection, while the constant reshuffling of hereditary material through sexual reproduction increases the frequency of the individual variations upon which selection can work.

There is thus a division of functions between adult and immature insect, the adult specialising on reproduction, on distribution, and evolutionary opportunity, while the immature insect remains more static and concentrates on feeding and growth. Sometimes the practical difference is slight, as in the cockroaches, where nymphs and adults live in the same situations, and eat the same food. Among grasshoppers – and particularly locusts – the mobility of the adult is used most effectively to carry large numbers of the insects to new feeding grounds and so to solve the problem of feeding very large populations.

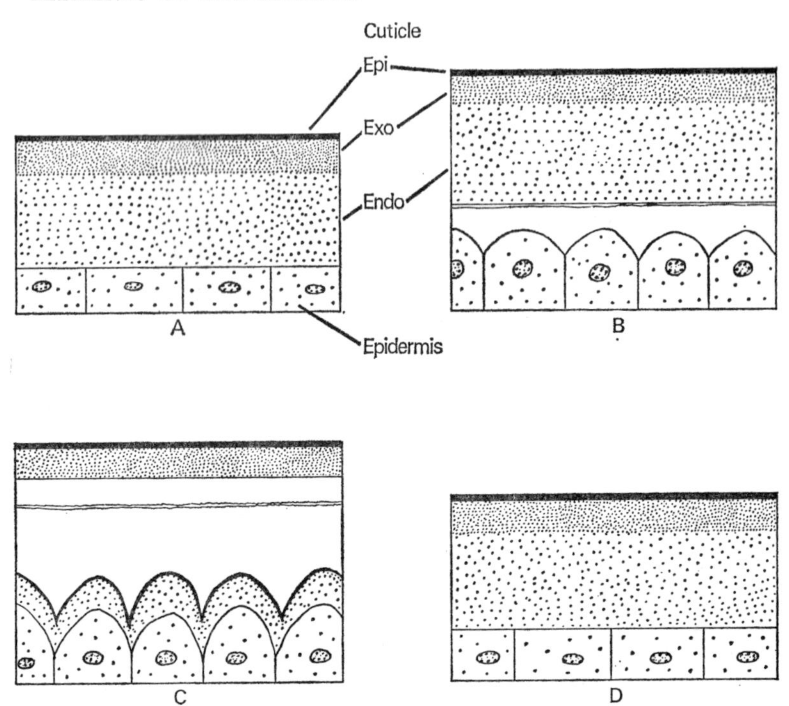

Figure 56. Successive changes in the form of the integument of an Arthopod during and after the moulting cycle (after Hinton).

The possibilities of a hemimetabolous life-history – i.e. of developing gradually towards adult structure – are exploited most fully by aphids, coccids and other related Homoptera. Adults as well as nymphs feed by sucking the sap of plants, and their food is virtually unlimited. It might have seemed that they could have just lived in large colonies, and not even bothered to fly. This is in fact what they do during the summer months, when they reproduce parthenogenetically, and devote all their energies to feeding and to building up the numbers of their populations. Yet even aphids have winged males and females in the winter, when they move to new hosts. This is an adaptation to suit climates in which the seasons are markedly different, and one of the important ways in which metamorphosis has been 'manipulated' by insects in such a way that they are enabled to live in the different climatic zones of the earth.

Corbet has suggested that such a use of metamorphosis has been made by Dragon-flies which were one of the earliest groups to emerge from the carboniferous forests. Corbet suggests that in the tropics Dragon-flies find it convenient to spend only a short time as a nymph, followed by a long life as a winged adult, especially in those parts of the tropics where there is a long dry season, with water for the aquatic nymphs. When Dragon-flies spread into temperate countries, where winters are too cold for the aerial life of the adult, they first overwintered in the egg stage. In summer there was then a rapid nymphal life, and then the adult appeared in time to lay eggs before next winter. Further north (or south) the summers become too short for this programme to be completed, and the nymph has had to adapt itself to survival through a long, cold winter, with the adult and its egg occupying the brief summer.

The flexibility of this sort of adaptation is shown by the Dragon-fly, *Anax imperator*, which lays eggs in early summer, and passes the winter as a nymph. Next summer metamorphosis continues, and those individuals that have reached the final instar by May emerge in that summer; those that have not progressed so far enter into diapause (Chapter 2) and pass a second winter as nymphs. The factor which induces diapause is said to be the progressive shortening of the days after the summer solstice.

Thus even hemimetabolous insects behave as if they lived two lives in succession, but the nymph of a hemimetabolous insect cannot be greatly different from the adult because it has to transform itself without the interpolation of an inactive pupal stage. Kennedy [54, 56] has suggested that the solitary and gregarious phases of locusts (see Chapter 20) may serve to overcome this difficulty, by a sharing of functions analogous to that between larvae and adult. In the solitary phase the locusts concentrate on feeding and building up a number of individuals – corresponding to the feeding and growth of the larval body – and then in the gregarious phase they concentrate on dispersal and the discovery of new feeding grounds, essentially functions of the adult stage.

It is perhaps surprising that the Holometabola have not made greater use of their advantage in this respect. Leaving aside for the moment those with aquatic larvae, other holometabolous insects have had three main evolutionary ideas: the most primitive of these is the predatory larva, with chewing mouthparts and three pairs of thoracic legs, examples being the voracious larvae of ground beetles (Carabidae), tiger bettles (Cicindelidae), and ladybirds (plate 23).

These are significantly called 'campodeiform larvae' because of their resemblance to *Campodea*, a very primitive insect of the Apterygota. It is surely no great evolutionary advance to spend most of one's life in the manner of such a poor relation.

The big advance in the evolution of the Holometabola was the eruciform larva, or caterpillar. This is a browsing type, designed to feed heavily and continuously on vegetable food that is there for the taking. This is typical of Lepidoptera as well as of their relatives, the Mecoptera, and it appears again, briefly, in the Saw-flies among Hymenoptera. Instead of taking the juices of plants by sucking in the manner of the Hemiptera, the caterpillar bites off and chews the plant tissues. To get enough sustenance the larva has to eat a lot, to eat quickly, and to be able to move steadily along as it eats its food away. In a close-range cine film it is frightening to watch this progress. These requirements account for the structure of the caterpillar, with its strong head and powerful mandibles; its segmented thoracic legs, helped by abdominal prolegs with grasping crochets; and its soft, extensible abdomen, by which it can shift its grip and pull itself along as well as accommodate its bulky food. An ugly, efficient feeding machine, replaced for a comparatively brief adult life by a very mobile, aerial insect, with large wings, and often only a puny body.

The third principal type of holometabolous larva is the apodous or legless type. Perhaps it is a mistake to classify this kind of larva as a 'type'. It is rather a convergent evolution among larvae which live within an abundance of food material, and have no need to come out into the open. They do not therefore need legs, either to seek their food or to escape their enemies, and they have gradually lost all trace of walking legs. This has happened, for example, among many beetles, where the complete transition exists from the active, campodeiform larva of primitive, carnivorous beetles, and of some others such as ladybirds (Coccinellidae); through the caterpillar-like larvae of some Chrysomelidae; then many fat, obese larvae of wood-boring and ground-living beetles, where the thoracic legs have become very short, or reduced to vestiges that have no useful function; and finally their series terminates in completely legless larvae like those of many weevils.

In the order Hymenoptera, the primitive Saw-flies (sub-order Symphyta) have larvae which resemble caterpillars, and feed openly on leaves, or burrow into stems. All other Hymenoptera (sub-order Apocrita) are without legs of any kind. The loss of legs, and in many cases the absence of any hard cuticle even in the head, are closely correlated with the sheltered and inactive life of the larva. Those of parasitic Hymenoptera hatch from their eggs in or on the food material, provided by the parent, and the larva has nothing to do except eat it. This is also true of the larvae of bees, wasps and ants, which either provide their larvae with enough food for life, or actively tend to their wants throughout larval life.

The true flies of the order Diptera are peculiar in having lost their larval legs very early in evolution, and so having had to get along without them in all their subsequent adaptations. It is remarkable how successful they have been in spite of this handicap. The larvae of many flies have developed *pseudopods* (false legs or prolegs), sometimes as well defined as the abdominal prolegs of caterpillars, but often rather indefinite swellings of the body-surface, armed with bristles or spines. For example, the maggots of House-flies and Blow-flies have on each segment a ring of irregular swellings, and a rash of small bristles like an unshaven chin, with the aid of which they can squeeze or wriggle through any kind of medium.

Unlike the apodous larvae, which are simplified for living in what Kennedy would call a 'soft' habitat, the legless maggots of flies are vigorously adapted to fend for themselves in any kind of environment, from the tissues of plants, through various kinds of decaying materials, to active parasitism in living animals. They are a deceptively simple kind of larva that is really a 'universal model'.

12
Adaptations to life in water

Though the invention of metamorphosis may be the most important adaptation of insects, the many adaptations to aquatic life are their most spectacular. Unlike the evolution of metamorphosis, which has followed a comparatively orderly course, the evolution of aquatic habits, and of the structures to go with them, has happened repeatedly and erratically throughout all the orders of winged insects.

Rather strangely, perhaps, it is pre-eminently the winged insects that have taken to the water. Among the Apterygota, *Petrobius maritimus* (Thysanura) occurs in numbers on the seashore, *Anurida maritima* (Collembola) even between tidemarks, where it is submerged by the tide, and the fresh-water collembolan, *Podura aquatica*, can move over the surface of water. These examples, however, are only marginal habitats of an essentially terrestrial group.

When winged insects appeared they soon turned to aquatic habits. Those two ancient groups, the Dragon-flies (Odonata) and the May-flies (Ephemeroptera) both spend the whole of their immaturity in water, except for the brief excursion of the subimago of the latter. These groups have persisted in the same way of life as their ancestors of Carboniferous times, and there is some fossil evidence that even at that remote time their aquatic nymphs were highly adapted. The adaptation must therefore have begun much further back in time.

We have seen that besides the two orders just mentioned there are three others that are aquatic in the same degree, i.e. they spend their immature stages in water, and then have an aerial life as adults. These orders are Plecoptera in the Hemimetabola, and

Neuroptera (partly) and Trichoptera in the Holometabola. Plecoptera and aquatic groups of Neuroptera, the Sialoidea stand near the base of their respective lines of evolution, and show certain affinities to each other; whereas the second half of the Neuroptera, the Planipennia are mainly terrestrial. Since these last have sucking mouthparts it would seem that they were a later evolution.

It seems possible, therefore, that the aquatic habits of Odonata, Ephemeroptera, Plecoptera and Neuroptera may be primitive, and derive from one common ancestor. The Caddis-flies, Trichoptera, are rather different. They appear as fossils comparatively late, in Jurassic or late Cretaceous times, and are nearly related to the essentially terrestrial Mecoptera and Lepidoptera. Even if they derived from moss-living Boreidae it still seems likely that they have returned to the water in what, in evolutionary terms, is comparatively recent times.

The other twenty orders are terrestrial in the sense that the normal habitat of both immature insects and adults is not in water. The term 'terrestrial' may be applied not only to those insects that live out on the surface of the ground, or on vegetation, in sand, soil or decaying wood, but those in rotting vegetable and animal material, some of it so moist that the problems of survival for the insect may resemble those of a truly aquatic life in deep water. Moreover most of these orders include families, genera or species that have taken up an aquatic life, and have done this so long ago that by this time they have adapted themselves structurally and physiologically to the rigours of life in the water.

The striking thing about aquatic insects is the way in which totally unrelated insects have adapted themselves for similar purposes in very similar ways. The special problems that insects have to face when they live in water are: (a) respiration (b) salt/water balance in the body (c) locomotion and (d) how the adult insects shall escape into the air.

(a) *Respiration in water*

Most insects, particularly nymphs and larvae, have a thin cuticle through which some oxygen can diffuse, and so they can keep alive under water, at least for a time, and if they do not move about actively. For an active and sustained aquatic life they have to rely upon the tracheal system, and this must be supplied with air. The operation of the tracheal system depends upon gaseous diffusion, and therefore if a submerged insect is to make use of its tracheae they must be filled with air, and not with water or any other liquid.

Obviously the first requirement is that the tracheae should not immediately fill with water when the insect is submerged. The cuticle of an insect is not easily wetted, and the spiracles which open into the tracheal system are often provided with glands that produce an oily, water-repellent secretion. Yet, even so, if an insect with open spiracles remains submerged indefinitely the air in its tracheal system will gradually shrink and the water creep in to take its place.

All insects therefore can survive for a time in water, and do not become waterlogged and drown as quickly as does a terrestrial mammal. By keeping to shallow water, and coming to the surface frequently, an insect can manage without any special breathing equipment, or a mere extension of the posterior spiracles. Larvae of many flies carry the posterior spiracles at the end of a *siphon*, an extension of the abdomen which may be fixed in length, or may be telescopic. A notable example is the very long siphon of the 'rat-tailed' larva of the Drone-fly, *Eristalis tenax*, and some of its near relations. This is strictly like the Schnorkel tube with which the bather walks about in water above his head, but unlike the human, the fly-larva does not immediately drown if it loses touch with the surface. The larvae of some Drone-flies crawl about in shallow water with their siphons reaching up to the air; they also swim in deeper water, with only occasionally, visits to the surface. This is also true of mosquito larvae, which lie at the surface for much of the time, taking in air, but periodically swim about below for short distances.

This method of breathing has been adopted independently by a number of different groups. The Giant Water Bugs of the tropics – Belostomatidae – are magnificent examples of the method, which they practise both as nymphs and as adults (plate 21). To be so surface-bound is dangerous, however. It is always easier to escape from one's enemies in a volume of space rather than over an area. Moreover organisms which lie for long periods at the surface are easily seen from below by fish and possibly by predatory insects such as the various water-bugs and water-beetles. There is an evolutionary advantage in the mobility that comes from being independent of the surface for much of the time, if not entirely.

One obvious way in which to continue to breathe while submerged is to take one's air down in the form of a bubble. The well-known water-beetles, *Dytiscus* and *Hydrophilus*, submerge with a bubble of air trapped beneath the elytra, or wing-cases, and in contact with the spiracles of the abdomen. A similar device, evolved quite independently, is found in the water-bugs, Naucoridae, Corixidae, and even Belostomatidae, although these last, as we have just seen, can reach to the surface with a tubular siphon. It is perhaps going too far to call this modification a 'device', since there is little structural adaptation except a hollowing of the upper surface of the abdomen, which increases the volume of the air-space.

In particular, there is nothing to keep the bubble from collapsing. The actual cause of the collapse is the gradual loss of nitrogen by dissolving irreversibly into the water, under the steady pressure of the surrounding water. While this is taking place, the insect is using oxygen from the bubble, thereby reducing the oxygen tension, and this is replaced by oxygen diffusing out from the water. Fortunately the rate of diffusion, or 'invasion coefficient' of oxygen is much higher than that of nitrogen, and it is estimated that during the time that a given volume of nitrogen in the bubble is being lost, about twice this volume of oxygen can be taken from the water and used by the insect. It is this fact that makes such a bubble function as a form of lung for the submerged insect.

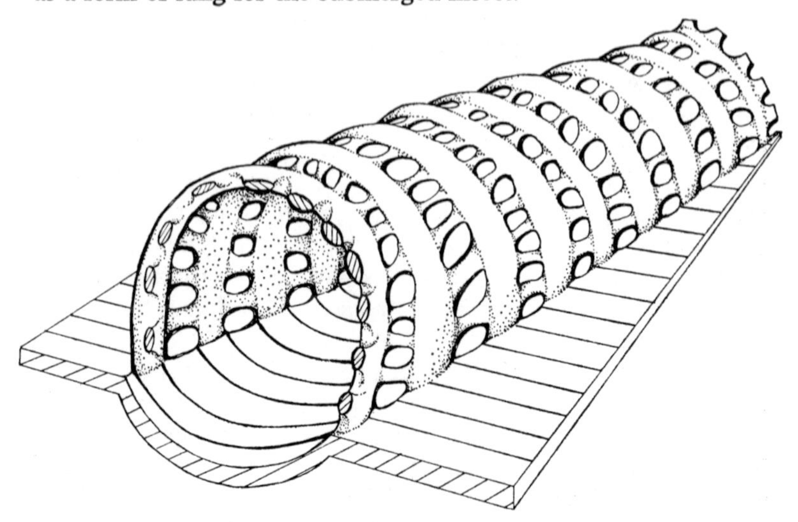

Figure 57. A 'plastron line' in the aquatic pupa of the crane-fly *Antocha* (after Hinton).

Even so, such a lung has only a limited life, and does not cater adequately for the respiration of a large insect. Thorpe [108] estimates that, indeed, very large insects such as the big *Dytiscus* do not get any substantial benefit from the air beneath their wings, though it may be enough for small Corixids to stay below much longer, especially if they are not very active.

The problem is to stop the bubble from shrinking, in some way to resist the pressure of the surrounding water. This is believed to be the function of the *plastron*, a thin film of air. Any greasy object dipped in water tends to take on a silvery appearance because a very thin layer of air adheres to its surface, being prevented from running together and escaping by surface tension. This effect is enhanced in many aquatic insects by the provision of a felt-like covering of very short hairs, regularly massed, and hydrofuge, or water-repellent on their surface. They trap a thin film that is very resistant to water-pressure; indeed it may resist up to four or five atmospheres. Hinton [44] has shown the same resistance to pressure in the film that forms when the eggs of certain flies are covered by rainwater, although in no imaginable circumstances could they have to resist such pressures in nature.

The mechanics of the plastron have been most fully studied in the water-bug *Aphelocheirus* (see Thorpe [108]), where the plastron was found to be rather more efficient in practice than was expected from theory. Notonectid bugs have hairy channels on the abdomen which act in this way, and similar devices occur sporadically in other orders. For example, an Ichneumon (Hymenoptera: Parasitica), *Agriotypus* parasitises the pupa of Caddis-flies, and pupates there inside its own cocoon. A ribbon-like thread extrudes from the cocoon, and acts as a sort of 'air-wick', replenishing the oxygen of the air trapped in the cocoon. By quite independent and remarkably convergent evolution a very similar device is used by the parasitic larva of the fly *Cryptochaetum* which lives in the scale-insect *Icerya*.

If an insect is to stay below indefinitely, without ever needing to come to the surface, it can prevent this shrinkage of its air-reservoir by having an entirely closed tracheal system with no open spiracles. This means, of course, that such an insect can never come out of the water and live a really terrestrial life, though it may be able to survive in a saturated environment: for example some larvae of psychodid flies, without open spiracles, live on surfaces that are

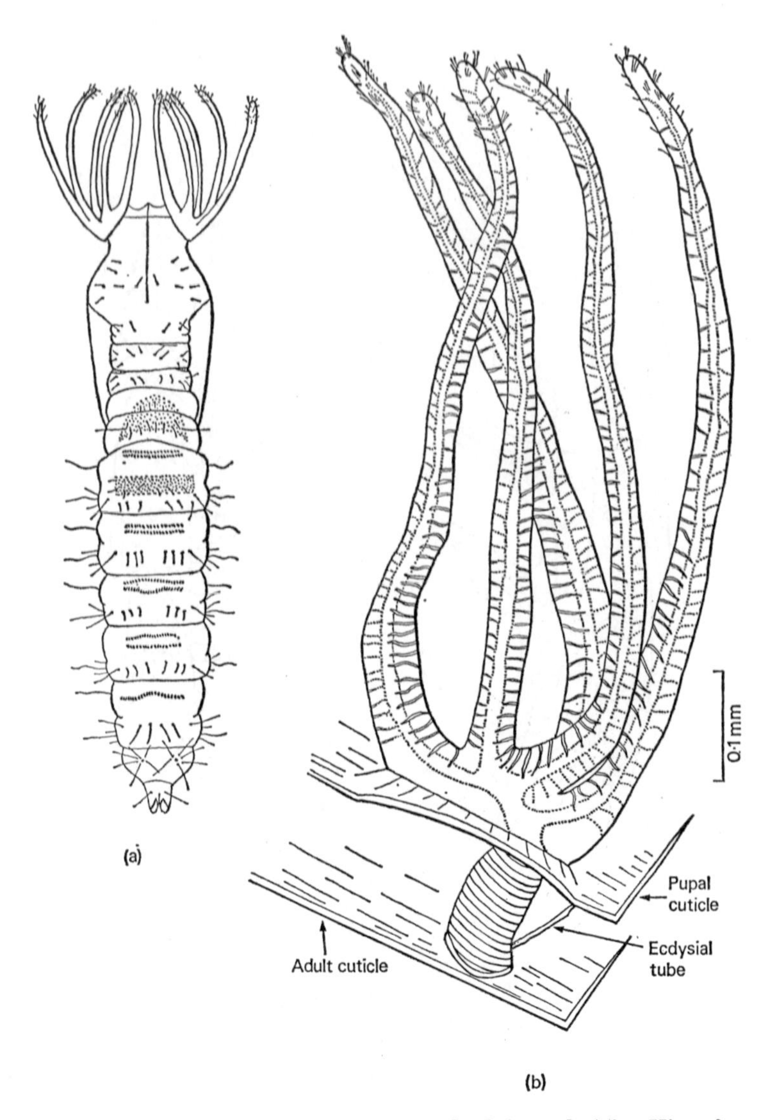

Figure 58. A spiracular gill in the crane-fly *Orimargula* (after Hinton).

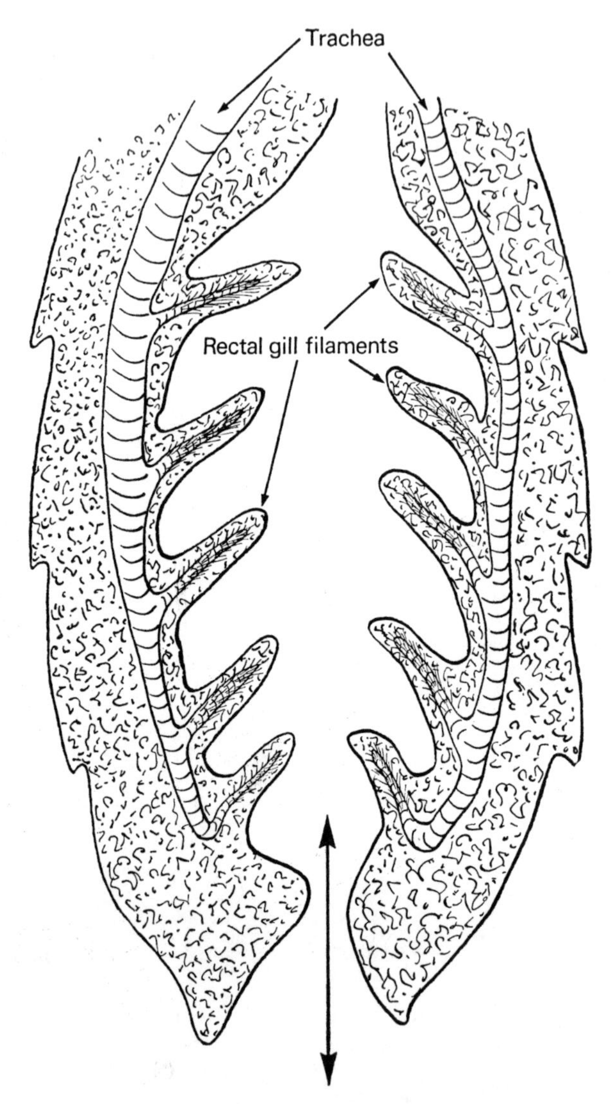

Figure 59. A rectal gill (after Wigglesworth).

perpetually wetted by spray from waterfalls. No adult insect has taken this extreme step in aquatic adaptation. It is a prerogative of nymphs and larvae, and it is a method particularly characteristic of the primitively aquatic groups of Ephemeroptera, Odonata and Plecoptera.

If the spiracles are merely closed, and the only supply of oxygen is by diffusion across this membrane, clearly the oxygen supply is going to be inadequate. All the nymphs just mentioned, which are relatively large, active creatures, have increased their supply by developing *tracheal gills*, extensions of the body-surface which provide a much larger area of membrane across which oxygen can diffuse into the tracheal system. Examples are shown in figures 59 and 64 and explained in the captions.

It is interesting to note the sequence in which these various devices appeared in evolution. A closed tracheal system with tracheal gills is found in all the orders whose immature stages are fully aquatic: Ephemeroptera, Odonata and Plecoptera among Hemimetabola; some Neuroptera (Sialoidea) and Trichoptera

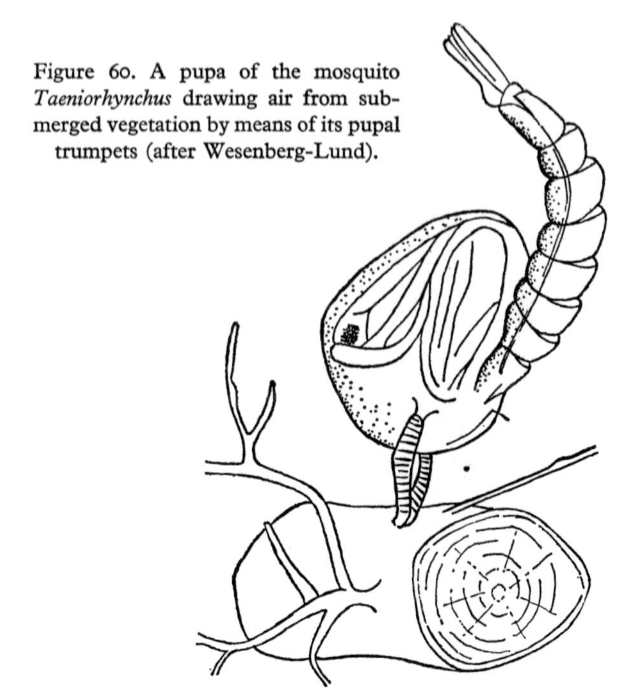

Figure 60. A pupa of the mosquito *Taeniorhynchus* drawing air from submerged vegetation by means of its pupal trumpets (after Wesenberg-Lund).

among Holometabola. Among the larvae of Diptera tracheal gills are not peculiar to those with a closed tracheal system, but on the contrary occur most often in larvae that also have open spiracles. In such cases – including certain bottom-living larvae of mosquitoes and chironomid midges – two different methods of obtaining air give the larvae a choice of living at the surface or at a depth.

In these more advanced orders of insects tracheal gills have evidently been developed independently. They are known in some lepidopterous larvae, and those of a number of aquatic beetles. Similarly the habits of creeping about and reaching the surface with a siphon, and that of taking down a plastron of air, have been 'invented' independently by quite different groups of insects.

13
Feeding habits 1: Plant-feeding insects

All living creatures must have food, which they use for two main purposes: as raw material for the construction of new tissues, and as fuel to be oxidised and thereby quickly release energy.

Plants vegetate, that is to say they stay in one place and grow. There are plants that are parasitic on other plants, and even some that are carnivorous, but in general plants are able to build up the complicated proteins of the cell-structure from quite simple chemical compounds, which they draw from the soil or from water.

Animals, in contrast, live at second hand. To make new tissues for themselves they need to start with comparatively complicated organic compounds, which they obtain by breaking down the proteins of either plants or other animals. Plants, as it were, make their own bricks, while animals use old bricks from some other structure.

There are three major types of feeding: taking proteins that have already begun to break down, as in vegetable or animal matter that is rotting or decomposing (*saprophagous* feeding); feeding on the cells of plants (*phytophagous*); and feeding on the tissues of animals (*carnivorous*). These are only categories of convenience, and in fact animals take food in any way they can get it. With our compulsion to classify things we try to distinguish subdivisions – for example a lion killing and then eating an antelope is *carnivorous*, whereas an ichneumon larva eating a caterpillar from inside and killing it is *parasitic*. These terms are useful as long as we remember that these nice distinctions mean nothing to a hungry animal.

Reverting to insects, it might be expected that the primitive method of feeding would be the saprophagous, taking the easily digestible products of vegetable or animal decay. This is not so. The primitive insects usually have mandibles, and chew their food (figures 19–21). The primitively wingless insects of the Apterygota may live on detritus, but this is only a name for tiny fragments of vegetable or animal tissues, not the liquid products of decay. Those Apterygota that have later developed sucking mouthparts, – Protura and some Collembola – are just the ones that have diverged from the insect line, and which some authorities would expel from the class Insecta.

The same is true of the primitive winged insects, which chew and masticate their food, whether animal or vegetable. Nymphs of Ephemeroptera mostly chew underwater plants, a few being carnivorous, and the adults not feeding at all. Odonata are carnivorous both as nymphs and as adults. Locusts, mantids, cockroaches, termites and Mallophaga among the primitive Hemimetabola all have chewing mouthparts, and it is only in the more recently evolved groups that the sucking of liquid food has been developed. Again, among Holometabola the chewing habit predominates among beetles, adults and larvae; it is fully established among Hymenoptera, and the sucking proboscis of bees is a later development; and chewing is a feature of primitive members of Neuroptera and of the larvae if not the adults of the Panorpoid complex, i.e. larvae of Mecoptera and Trichoptera, the caterpillars of Lepidoptera, and the primitive larvae of Diptera. Biting mouthparts are retained by some adults of this group, notably adult Mecoptera and the primitive family Micropterygidae, formerly included in Lepidoptera, but now placed in a separate order Zeugloptera.

Clearly biting and chewing is the ancestral habit in insects. Sucking is a later evolution, both in adult insects and in nymphs and larvae.

Detritus-feeders

Detritus is the name given to the accumulation of small fragments produced when something is rubbed away. The material is not changed in nature, but only reduced to small particles. Detritus-feeding insects eat and digest the original proteins of their vegetable and animal food, whereas saprophagous insects cannot use this material as food until it has been transformed into simpler compounds by decomposition or decay.

Organic detritus is present wherever living things exist, but the problem is to collect enough of it to subsist upon. On land, at least, detritus would seem to be a diet suited to small, active insects, and indeed the tiny, active Apterygota are believed to feed in just this way. An example is the fire brat (plate 2) (Thysanura) running about near the hearth. But these creatures do damage to books and papers, and it is evident that they are not addicted to tiny particles as such, but will eat anything that is in a condition in which they are able to deal with it.

These and other insects that are said to feed on small particles, and hence are classed as detritus-feeders, generally flourish in damp places. To some extent this is a physiological necessity for very small insects, which otherwise would lose too much moisture from their relatively large surface. Damp places, however, are favourable to the development of moulds and other forms of minute vegetable growth, and it may often be that these are the real food of the insects.

Among bigger insects, detritus-feeding becomes scavenging. The most successful insect scavengers are the cockroaches (plate 13), which are usually described as *omnivorous*. The domestic cockroaches are the only ones whose diet is well-known, from their depredations in the kitchen, and from their tastes in the breeding-cage. They appear to masticate anything of either animal or vegetable origin that comes their way. Less is known about the wild cockroaches, but as they live under leaves and among general refuse, it is likely that they, too, are omnivorous.

The remarkable thing about cockroaches is that they are an ancient group, and have survived with little change since early Carboniferous times. It seems, therefore, that general feeding by chewing with the mandibles provides a good diet, which has kept the cockroaches alive for something like three hundred million years. Why did insects ever bother to do anything else? In all the millenia during which the more specialised groups of insects have slowly evolved, and developed elaborate sets of mouthparts for peculiar diets, cockroaches have continued in the same old way. The Darwinian idea of the struggle for existence, and the survival of the fittest, seems absurdly dramatic when we think of the cockroaches.

Yet even here there is the beginning of an urge which constantly appears throughout the evolution of insects, the urge to concentrate on animal protein as a more concentrated source of food than a vegetarian diet. The earwigs (Dermaptera), close relatives of the cockroaches, and also general feeders, have a more pronounced tendency towards animal food, and in captivity will take living insects. No one knows if earwigs are carnivorous in a free life, but the mantids – which are placed close to the cockroaches, even in modern classifications, are rapacious feeders on other insects.

Reverting, however, to detritus-feeding and scavenging on natural organic fragments, we find these habits in Embioptera and Psocoptera, the latter especially being very destructive to books, wall-paper, and natural history specimens. Mallophaga, the biting lice of birds, are specialised scavengers, eating the debris that accumulates on the surface of the skin at the base of the feathers.

Among the larger and more highly evolved orders of insects, detritus-feeding is less common in the adult insects. It is perhaps retained most in the beetles, a conservative order, as we have already noted. Ants are generally scavengers, but this is not so much a retention of an old habit as a part of the elaborate system of social life that ants have evolved. Similar evolutionary trends have been followed by ants and by termites (Isoptera), although these two orders are far apart in the plan of insect evolution (chapters 15 and 17).

Elsewhere among the 'higher' insects, detritus-feeding is mainly a larval habit. The outstanding scavengers among terrestrial larvae are those of the fleas (Siphonaptera). The adult fleas take no food other than blood, and spend all their adult lives on a vertebrate host. The larval fleas do not live on the host, but in crevices round about the nest, sleeping-place or habitation of the host. There they feed on detritus of animal origin, scraps from the skin, dried blood, and so on. Other scavenging terrestrial larvae occur sporadically in different families of the various orders of Holometabola, but the most highly organised detritus-feeders among larvae are the aquatic ones. Small particles suspended in water are the easiest form of detritus to feed upon, because the insect can stay in place and filter out the particles as they go by. On the other hand this requires the evolution of special filtering apparatus, and therefore we find that aquatic filter-feeding larvae belong to relatively advanced groups of Holometabola and, moreover, that they have evolved from primitive chewing types. Since this is an advanced form of feeding we will leave it for the moment, and return to primitive mandibulate chewing insects.

Plant-feeders

The mouthparts of a cockroach (figure 19) and those of a locust are examples of generalised equipment, suitable for masticating any kind of solid food. The cockroach is a general feeder, and uses its mouthparts to tackle anything that is not too large or too hard to bite and chew. The locust, on the other hand, uses essentially similar equipment mainly for the purpose of eating from a living plant.

If you have seen a film of a locust feeding, or better still a living specimen, you will have a vivid picture of the efficient way in which the insect cuts slice after slice from a leaf with its mandibles, manipulates the pieces with its flexible maxillae and labium and their palps, and packs the pieces of leaf into its pharynx as fast as they are cut. This performance illustrates some of the basic problems of feeding upon growing vegetation. Plant-cells have walls of cellulose which must be pierced or broken to reach the fluid contents which are the real source of nourishment to the insect. Few insects can digest cellulose until it has first been broken down chemically into simpler constituents. Even termites, many of which are remarkable for their ability to eat wood, depend mainly upon having the cellulose first pre-digested for them by colonies of Protozoa, unicellular animals which live in the beginning of the hind-gut of the termite.

For most insects, as for man, the cellulose framework of plant-tissue is mere roughage, and the biting off of fragments must be followed by chewing or mastication to smash up the cell-walls. Herbivorous vertebrates either have teeth with elaborate grinding surfaces like those of horses, or bring their food back again for further chewing as do cows and other ruminants. It is remarkable that browsing and grazing insects can masticate their food sufficiently with a single pair of mandibles, but perhaps the problem is less difficult because the insect itself is much smaller. The mandibles of a grasshopper are much more instruments of precision for crushing the walls of individual plant-cells than are the great teeth of a horse.

Plant-food is outstanding for its bulk. Since the cellulose is mostly waste, any herbivorous animal must pass a great bulk of vegetable matter through its intestine in order to get enough digestible material for its needs. This is one reason why locusts eat so quickly; they have a lot to get through. Another reason, peculiar to locusts and other insects that feed in huge swarms, is that they may completely strip the vegetation of its foliage, and if the total supply of food is thus limited, those insects that eat most quickly will get the most food.

A long way in evolution from locusts, the caterpillars of butterflies and moths show the same responses to the same problems. Caterpillars, too, eat quickly and in great bulk. They, too, can strip a bush of its foliage if they are numerous enough, and those caterpillars that eat too slowly will be underfed. Most of them will get enough to be able to complete their metamorphosis, but the underfed ones will give rise to undersized adults, which may also be sexually immature or even intersexed.

Both locusts and caterpillars have an abdomen with a soft cuticle, and capable of becoming swollen and distended with bulky food. Either the sac-like crop or the mid-gut itself is capacious, so that the insect can take in a mass of food when it is available, and digest it over a period.

Relatively few of the insects living today have remained content with biting and chewing living plant-tissue. The outstanding examples are the groups just mentioned, the Acridids (locusts and grasshoppers) on the one hand, and the caterpillars of Lepidoptera on the other. Among relatives of the locusts, the cockroaches and crickets are omnivorous scavengers; the long-horned grasshoppers and bush-crickets (Tettigoniidae) are partly herbivorous, but include some that take any food, and some that have become carnivores. The stick-insects (Phasmida) are plant-feeders with large appetites – some of them are among the biggest of insects, reaching to over a foot in length.

The damage done by foliage-eating insects will be discussed in Chapter 20.

Sap-sucking insects

A more efficient way of extracting the juices of plants is by sucking. A living plant in the growing season has a steady flow of sap, a highly nutritive fluid, and many insects have modified the original masticating type of mouthparts into a hollow tube which can be pushed into the plant, and through which the sap can be withdrawn.

The sucking-tube is always double because saliva plays a large part in the feeding of sucking insects. Chewing insects, of course, have salivary glands, and use saliva to lubricate the many parts, and to mix with the food to start off the process of digestion, as we do ourselves. But to sucking insects, saliva is even more vital. The sucking channel is narrow and long, and is easily blocked during use. The contents of plant and animal cells are of varying consistency, and include many particles that are too big for the narrow sucking-tube. An up-and-down movement of saliva is necessary to flush the mouthparts, and a certain amount of pre-digestion takes place by expelling saliva into the tissues and drawing it up again, together with part of the cell-contents. These functions of the saliva are even more important in blood-sucking insects than they are in plant-feeders.

The great order of plant-sucking insects, is of course, the 'bugs' of the order Hemiptera (Heteroptera+Homoptera). These have abandoned the use of chewing mandibles, and instead have the mandibles and maxillae in the form of lancets or stylets, lying in a groove in the upper surface of the labium, which forms a *proboscis* or *rostrum*. It is remarkable that a very similar type of sucking proboscis has been developed by the true flies of the order Diptera, which are Holometabola, and far removed in evolution from Hemiptera. Moreover, the flies use that proboscis to suck up a wide range of fluid foods, but they do not pierce and suck plants. This similar development of the proboscis in two remote orders of insects is one of the most remarkable examples of convergent evolution among insects.

Hemiptera are grouped into two sub-orders, Homoptera and Heteroptera, and Homoptera fall again into two sections, Auchenorrhyncha and Sternorrhyncha. The Homoptera-Auchenorrhyncha are most small, rather active, soft-bodied insects like the froghoppers (Cercopidae), the leaf-hoppers (Jassidae) and the treehoppers (Membracidae). The recurrence of the name 'hopper' shows that these are mobile insects, which catch the eye by their activity. It seems likely that the primitive Hemiptera, the ancestral bugs, were insects such as these, which fed by sucking plant juices, and escaped danger by leaping blindly away into the unknown. Cicadas and lantern-flies (see page 56) are rather unexpected offshoots of this group, but the two main stems of evolution among Hemiptera have been to Heteroptera on the one hand and to Homoptera-Sternorrhyncha on the other. In their relation to plant-feeding these are contrasting groups.

Heteroptera have continued and exaggerated the mobility of their ancestors. They are the plant-bugs that appear as individuals on the vegetation, the nymphs walking about freely, and the adults taking off to fly to another plant in a seemingly purposeful way. Most Heteroptera feed upon the juices of plants, but Reduviidae and a few members of other families suck the blood of vertebrates, while most of the water-bugs catch and suck other insects and small aquatic animals. The process of piercing and sucking another insect or a vertebrate animal is not essentially different from piercing and sucking a plant and the sharp, stylet-like mandibles and maxillae can be made to serve the same purpose.

From the ancestral Auchenorrhyncha, the Homoptera evolved in another direction into the present day Sternorrhyncha, the aphids and their relatives. These are distinguished by their static habits, and spend a great part of their life packed close together on a stem or a leaf, often so close together that the plant may be said to be encrusted (plate 42).

The effects on the plant of the activities of sap-sucking bugs are threefold. The most important in practice is the spread of plant-diseases, especially those caused by a virus or by bacteria. Then there is the loss of sap, which is trivial where one insect is concerned, but becomes serious because of the immense numbers that may be involved. Thirdly, there is the hindrance to the normal growth and respiration of the plant that results from the clogging effect of an incrustation of insects and from the wax and honey-dew that they produce. Honeydew is a sugary residue passed out through the anus. Wax is a by-product of the excretion of the insects, coming from wax-glands in the cuticle, including the conspicuous *cornicles* of aphids (plate 22). Originally, no doubt, this was a way of getting rid of waste-products, but many groups of Hemiptera have made a positive use of it as a protection. The froth of fine waxy filaments that covers the Greenhouse White-fly (Aleurodidae), the Woolly Aphis of apples, and some of the Lantern-flies, is used in this way. The shellac of scale-insects (Chapter 21) is a more elaborate waste-product of this type.

Feeders on nectar and pollen

Sucking mouthparts need not be used to pierce the plant-tissues, since both protein and carbohydrate food material can be obtained from flowers without piercing. Pollen, which includes the germ-cells of the plant, contains vegetable protein. Nectar is a sugary solution produced by the plant as an additional attractant to pollinating insects. Obviously insects can have evolved the habit of visiting flowers for food, and any special mouthparts that may be needed, only since the flowering plants first appeared in mid-Cretaceous times. Three of the most advanced orders of insects, Lepidoptera, Hymenoptera and Diptera, have flourished in that period, and in each case their evolution can be closely linked with that of the flowering plants themselves.

Butterflies

We all associate butterflies and moths with flowers. Apparently the earliest ancestors of the Lepidoptera had biting mouthparts, but today the only survivors of these are the family Micropterygidae, the order Zeugloptera of some classifications. For practical purposes all Lepidoptera that have mouthparts at all have a sucking proboscis, but this is differently developed from that of the plant-bugs. In Lepidoptera (plate 39) there are usually no mandibles, and the labium is reduced to a small plate. The long, coiled proboscis so familiar to us in butterflies consists of the outer lobes or *galeae* of the two maxillae, which are locked together by a series of hooks to form a sucking tube. This type of proboscis is obviously an extreme development permitting the habit of probing deep into the recesses of flowers to reach the nectar, which is usually so placed that the insect cannot reach it without also brushing against the pollen.

We like to think of butterflies as beautiful creatures, their colours mingling with the colours of the flowers to form a picture of warmth and gaiety. There is no evidence that the butterflies themselves see their world like this, and they will drink just as contentedly by sucking up moisture from a stagnant pool, or water in a rut in the road, from moist dung, or on sweat from a living animal. In fact one collector said that he often deliberately threw away his chewing-gum at intervals along the side of a path, and returned later to take the butterflies that had come to suck at it. Juices of overripe fruit are also attractive to butterflies, which sometimes can even pierce the rind by means of teeth on the proboscis.

It should also be remembered that butterflies are only a minority of Lepidoptera, and among moths there is much more variation in the level of evolution of mouthparts. Some moths still have traces of mandibles; some have the galeae rather poorly developed into a proboscis; and some have the mouthparts so greatly reduced that they evidently do not feed at all in the adult stage (plates 49, 50).

Bees

Association with flowering plants has also left its mark on the evolution of Hymenoptera. The adults of this order are mandibulate insects and in exact reverse of the Lepidoptera they have kept their mandibles even when they have also developed a sucking proboscis. This is the order of the great families of *social insects* (see Chapters 16–17), and among these the mouthparts are put to a great many uses besides eating. Watch an ant staggering along with a burden bigger than itself, and the mandibles are then seen as a powerful tool for carrying not only food, but nest-building materials as well. Ants often pick up and move their larvae and pupae by carrying them in the mandibles (plate 46).

It is the bees among Hymenoptera that have developed the practice of feeding from nectar, in place of the carnivorous diet of the wasps, and the general feeding of the ants. The proboscis of a bee has some superficial resemblance to that of a butterfly, but it is formed quite differently from the primitive mouthparts; not the outer lobes of the maxillae, but the inner lobes, or *glossae* of the labium (figure 21). Like the proboscis of the butterflies, that of the bees can be unrolled by blood-pressure, and withdrawn by muscular action.

Flies

Among true flies of the order Diptera the picture is still more complex. No adult fly has chewing mandibles, and though a few have acquired the ability to crush small insects, or to rasp away the skin of other animals, they have had to evolve new types of tooth-like attachments for this purpose. The great majority of adult flies are suckers. Some pierce as well, but they do not pierce plants, and so they will be considered in Chapter 19 with other carnivorous and blood-sucking insects. Under the present heading need be noted only that the piercing proboscis of a fly – or at least of one of the more primitive blood-sucking flies such as mosquitoes and biting midges – is of very similar construction to the plant-piercing proboscis of the Hemiptera. The mandibles and maxillae are drawn out into pointed stylets, and ensheathed in the labium, though this latter in Diptera is not divided into segments.

Only a minority of flies use their proboscis for piercing, and these not all the time. Nearly all adult flies suck liquids in some form: water, nectar, sugary exudations from stems and fruits, sap from wounded trees, as well as liquids of animal origin such as perspiration, mucus, pus from wounds and the liquids from excreta and rotting animal tissues. Though only the fluids of the first group come into the category of 'plant-feeding', the process of sucking is the same for all of them.

The sucking organ of the flies is the *labium*, which is functionally divided into two parts. The stem has a dorsal groove, which is roofed over by the labrum, and in which lie the mandibles and maxillae (if any) and the hypopharynx with its salivary duct. The labial groove forms a food-channel along which fluids are sucked into the pharynx. At the tip of the labium is a pair of lobes called the *labella*, which look like a pair of sponges, and are used as such. Each labella has a group of branching channels or grooves called *pseudotracheae*, because their walls are supported by rings of chitin, which give them rather the appearance of the true tracheae of the respiratory system, except that the walls are incomplete. This means that when the labella are pressed against a mass of fluid or semi-fluid food material this can be drawn into the labella as into a sponge.

We have all seen a House-fly feed from a lump of sugar in this way. In many flies that feed from exposed surfaces the labella are relatively large. On the other hand, in flies that use their proboscis to search for their food in the interstices of flowers the stem of the labium is long, sometimes very long as in the Bee-flies (family Bombyliidae). Then the labella are narrower and pointed, and the pseudotracheae are reduced in size or lost altogether.

Leaf-miners and Stem-miners

So far we have dealt mainly with the feeding of adult insects, or of the nymphs of Hemimetabola, which live in much the same way as their parents. The larvae of Holometabola also attack plants, and as the larva is often the principal feeding stage of these insects, when enough food is taken to supply the adult as well, their activities deserve attention.

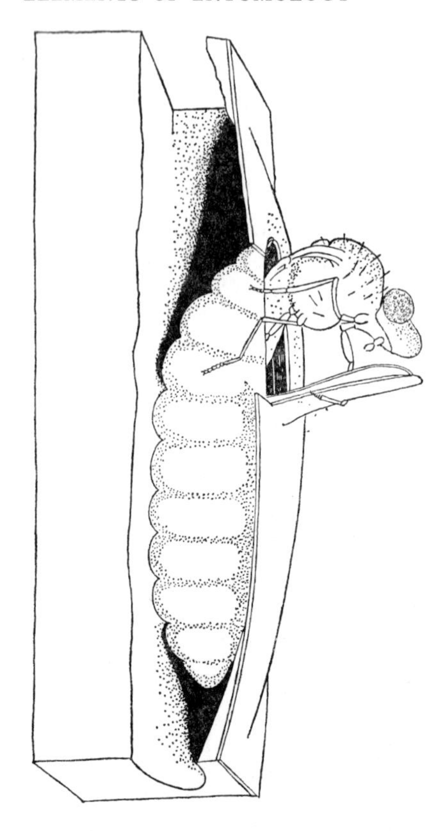

Figure 61. Leaf-mining. The larva (of an agromyzid fly), has fed on the parenchyma of the leaf, and formed a pupa beneath the epidermis; now the adult fly is forcing its way out, using its *ptilinum*, the inflatable sac protruding from the head (after Hering).

Mandibulate larvae which are surface feeders, notably the caterpillars, merely chew away the exposed plant tissue like any other chewing insects, but a number of other larvae have become internal feeders. Sometimes the egg is pushed through the epidermis of the plant by the ovipositor of the female; sometimes the eggs are laid on the surface and the newly hatched larvae, still very small, bore into the tissues. In either case the larvae begin to feed from the inner tissues of the plant leaving the surface intact, and thereby making a tunnel or *mine*.

Mines are most conspicuous when they are made in the leaves, and such *leaf-mines* can often be seen on either deciduous plants or evergreens. The larva eats away the parenchyma of the leaf as shown in figure 61, leaving a thin layer of epidermis both above and below. Since this is thin the mine appears yellow in colour, and shows up clearly in contrast to the deeper green of the untouched areas. A familiar example is the mines made in leaves of the holly by the Agromyzid-fly, *Phytomyza ilicis*.

Leaf-mines are of two kinds, blotch mines and serpentine mines. Blotch mines are made by larvae which eat about in all directions from one position, and so create an irregular blotch. Serpentine mines are the work of larvae which move forward as they feed, and so make narrow tunnels, which often follow along the ribs and veins of the leaf. Some leaf-mining larvae pupate inside the leaf, and then the adult insect has to break out. Other larvae leave the leaf when they are fully fed and fall to the ground, where they pupate in the soil.

Leaf-mining is a habit that has evolved independently in a number of different families in all four big orders of Holometabola, Lepidoptera, Diptera, Hymenoptera and Coleoptera. Among Hymenoptera this is mainly a habit of the larvae of Saw-flies, which in fact derive their common name from the saw-like ovipositor of the adult female which is of a suitable shape to push the eggs into the tissues of a plant. Most Saw-fly larvae are like caterpillars, and feed on the surface of plants, but some penetrate beneath the surface, and those of one family – Cephidae, the Stem-saw-flies – damage cereals and other plants by tunnelling in the stems.

Among beetles it is mainly larvae of three families, Chrysomelidae, Buprestidae and Curculionidae that are concerned in mining. The last family are the notorious weevils, snout-nosed beetles of characteristic appearance (plate 30). The mandibles of the adult are carried at the tip of the narrow snout, which is used to bite a hole into grain, nuts, fruits, stems or leaves. It is said that the eggs are sometimes laid in this hole, but in any event the larvae tunnel about in the tissues and do a great deal of damage. Weevils are among the most serious pests of plants, and especially of plant-products in store.

Among flies the family Agromyzidae consists mainly of flies with leaf-mining larvae, though others occur sporadically in other families; for example *Scaptomyza* among the Drosophilidae, a family of small fruit-flies the larvae of which live mostly in fermenting vegetable matter, overripe fruit and so on. The greatest number of leaf-mining species of any order occurs among Lepidoptera, and there are said to be about four hundred species in the United States alone.

There are no clear divisions, either structurally or physiologically, between leaf-mining insects and their near-relatives which feed superficially on the surface of the plant. Leaf-mining larvae usually have mandibles, but those of Agromyzid and Drosophilid flies are true maggots with mouth-hooks, like the maggots of House-flies and Blow-flies (figure 72c). It is possible to distinguish the mouth-hooks of plant-feeding maggots from those of maggots that feed in soft tissues or decaying materials. The plant-feeders often have ridges on the mouth-hooks, as adaptive modifications to suit the harder food material, which must be hacked away, as with a miniature pick.

The pattern of a leaf-mine is often highly characteristic both of the insect concerned and of the plant involved. Indeed there has been much classification of leaf-miners by their mines rather than by their own structure. This is an unsafe procedure, because clearly the plant contributes to the mine as much as the insect. This is particularly true of serpentine mines, which are the most characteristic because the larva tends to follow the easiest routes, and these are largely determined by the skeleton structure of the leaf.

Wood-borers

While discussing mines we ought to consider borers. These have a similar mode of attack, but are usually described as boring rather than mining when they attack a large plant such as a growing tree, or a mass of wood.

The notorious borers are the workers of termites on the one hand, and the larvae of certain families of beetles on the other. Among Hymenoptera certain large Saw-flies, such as *Sirex* (or *Urocerus*) *gigas*, the giant wood wasp, or horntail have wood-boring larvae. They get their common name because the adult insect is one to two inches long, and like a wasp in appearance, usually with a long sting-like ovipositor. With this powerful instrument the insect bores through the bark of trees and lays its eggs in the hole and the larvae bore into the heartwood, and may do much local damage. Among Diptera the larvae of the archaic family Pantophthalmidae are large, and powerfully equipped with mandibles with which they can bore through living wood. They are sometimes a pest, but fortunately are restricted to the forest areas of tropical South America (plate 7).

Termites as social insects will be discussed more fully in Chapter

15. As wood-borers they fall into two categories, the subterranean termites and the dry-wood termites. The subterranean termites live in soil and feed upon moist wood only. Their natural habit is to break up buried wood, or dead stumps projecting from the soil. They will equally well attack artificial wood structures such as wooden houses, if these are in contact with the soil. To keep them out of buildings these must be raised on brick or concrete supports, well clear of contact with the soil. Even then, the termites may reach the wood by constructing a tunnel across the intervening surface. By this means the galleries in the wood are kept full of moist air from the soil, and thus maintain an atmosphere suitable for the termites to work in. Termites of other species are able to live and feed in dry wood, and hence do not need to maintain contact with the soil.

Termites are thought of primarily as wood-boring insects, because this activity of theirs is destructive to human property. In fact they are an omnivorous group of primitive origin, but highly evolved in certain directions, notably in social organisation. The ability to digest wood with the help of intestinal Protozoa (see page 49) is a refinement. There are all grades of evolution, from purely soil-living types feeding on humus, moulds and fungi, to the dry-wood termites already mentioned. Probably the wood-feeding habit arose from soil-living termites which entered soft, decaying wood, but it is also considered that some of the most advanced termites have given up their intestinal protozoa and reverted to a diet of humus and fungus.

The *beetles*, the other great group of wood-boring insects, are also chewing insects, equipped with mandibles both as adults and as larvae. Beetles of one sort or another feed on almost all kinds of digestible material, and, as in termites, wood-feeding is only an extreme evolution in one direction.

Thus, the stag beetles (Lucanidae), those big, heavily armoured beetles that appear in one's garden on a summer's evening, have large, fleshy larvae with strong mandibles, with which they feed in rotting stumps. They can be a great help in weakening a tree-stump to the point at which it can easily be uprooted. Larvae of Buprestidae are vegetable feeders, many of which make tunnels beneath the bark of trees. The best known borers are members of the family Anobiidae, to which belong the furniture beetles, *Anobium punctatum*, and the Death-Watch Beetle, *Xestobium rufovillosum*.

These two make a network of burrows in wood, both indoors and out, and weaken it until it is a mere crumbling mass. Yet other members of the same family of beetles devour rather dry material of all kinds, without making burrows. Closely related to them is *Ptinus tectus* and other Ptinidae, which damage dry materials in store and are a menace to museum collections. The Powder Post beetles of the families Bostrychidae and Lyctidae go furthest, living on dry and seasoned wood and reducing it to a fine powder.

Some wood-boring beetles are said to have the help of Protozoa to digest the cellulose, in the same way as the termites, but the habit of wood-boring is more sporadic among beetles, and evidently many beetles have no special adaptation to this diet. They merely make use of their powerful mandibles to crush the cell-walls thoroughly and then they feed upon the sugary contents of the cells. The household wood-borers each favour wood of a particular age and degree of seasoning, its suitable condition being determined by moisture content, fungoid growth and the changes in its sugar-content that take place as time goes by.

14
Feeding habits 2: Insects taking animal food

The main advantage of animal food over vegetable food is that it is more concentrated. Thinking again of the land-vertebrates, the herbivores graze almost continuously and pass an enormous bulk of food through their intestines, whereas the carnivores feed in definite meals, often with long intervals between them.

This is equally true of insects. Those which browse on foliage – for example the locusts and grasshoppers on the one hand, and the caterpillars of Lepidoptera on the other – feed continuously or at any rate for long periods of time. During this time, when they are preoccupied with their feeding, they are vulnerable to attack by predators. Carnivorous insects on the other hand are generally more mobile, taking a meal whenever the opportunity arises.

In the broad sense an insect is *carnivorous* if it lives wholly or mainly on food of animal origin. If it catches another animal – for example another insect – kills and eats it, then the attacker is said to be *predaceous*, or a *predator*. In such cases the victim is usually smaller than the attacker, or at least not very much bigger. If an insect attacks an animal very much bigger than itself, for example a vertebrate animal, the attacker is said to be a *bloodsucker*, or a *parasite*, or both. To be classed as a parasite an insect must normally remain on or in its host animal for some time. Thus a Tsetse-fly or a mosquito, which visits its host for a meal of blood and then flies away to digest its meal is a bloodsucker, but not a parasite. A flea or a sucking louse is a parasite because it remains with its host almost continuously.

Any kind of mouthparts that are suitable for taking vegetable food can be used to eat animal food, and so there is no special kind of equipment that is evolved only by carnivorous insects. It is true that many predaceous insects have evolved special structures for capturing and holding their prey, usually by adapting parts of the legs as prehensile organs: examples are the fore-legs of mantids (figure 24) and those of certain robber-flies (Asilidae). This is only one of many ways in which the legs have been modified to suit the life of the insect. Moreover a few robber-flies have the fore-legs developed in much the same way as the mantids, an order of insects very far removed from robber-flies in the evolutionary scale; while at the same time other robber-flies, very closely related and of very similar habits, have no modifications of the legs, yet manage to catch and eat other insects just as well. It is another example of the way in which evolution has given some insects complicated structures to accomplish what other insects do with no trouble at all.

The habit of taking food of animal origin has arisen many times during the evolution of insects, and, like the habit of living in water, it occurs at random in many different orders. The only large order to be entirely carnivorous is the Odonata, and they do it thoroughly. Both as nymphs and as adults they are predaceous, that

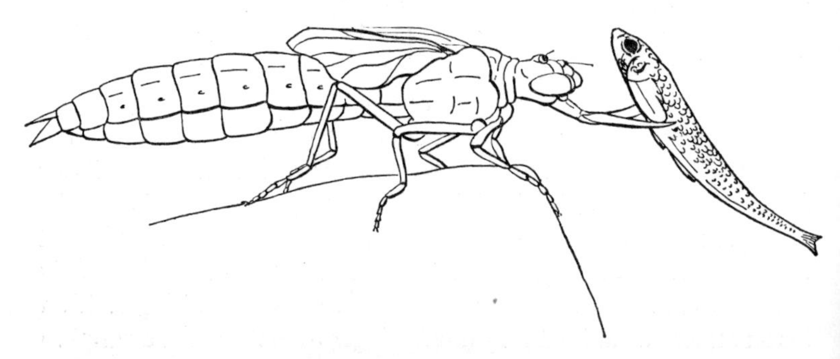

Figure 62. A dragonfly nymph (order Odonata) captures a small fish by shooting out its 'mask'.

is they actively catch other insects, or other small animals, and devour them. For this purpose they use mandibulate mouthparts not far removed from the primitive type. The principle modification is in the development of the labium, which has hooks rather like a second pair of mandibles. These obviously enhance the Dragon-fly's power of seizing its prey, and in the Dragon-fly nymph the labium is more elaborately modified into the *mask*

1 A wood-louse (*Crustacea: Isopoda*) rolling over, and showing its succession of similar segments, each with a pair of jointed appendages, here developed as walking legs.

2 The Fire Brat, *Thermobia*. In this primitive insect the body segments are already divided into head, thorax and abdomen, but the abdomen retains vestiges of its segmental appendages.

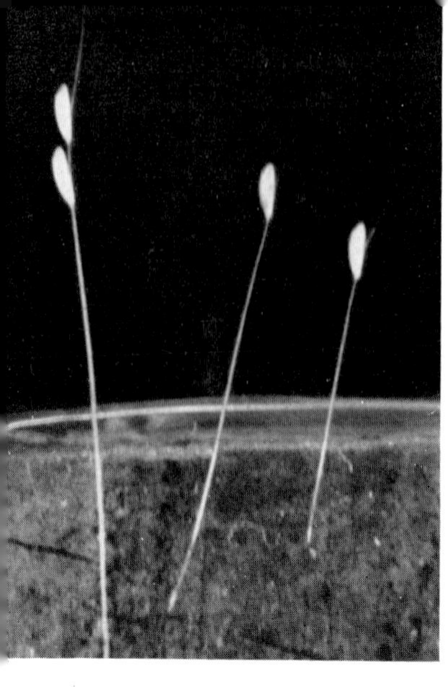

3 (*left*) Hairs from the leg of a horse, with attached eggs of the horse bot-fly, *Gasterophilus intestinalis*.

4 (*centre*) A beetle larva, with three pairs of walking legs on the thorax, and no abdominal appendages.

5 (*bottom*) Aphid-feeding larva of one of the hover-flies (Syrphidae): a flattened crawler, with no appendages except for the twin-spiracles posteriorly (*left*).

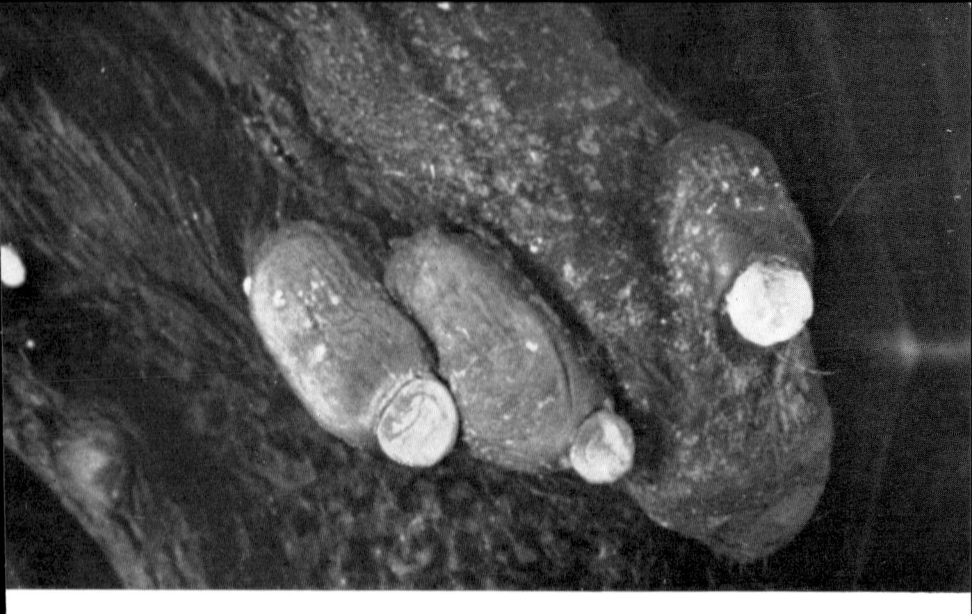

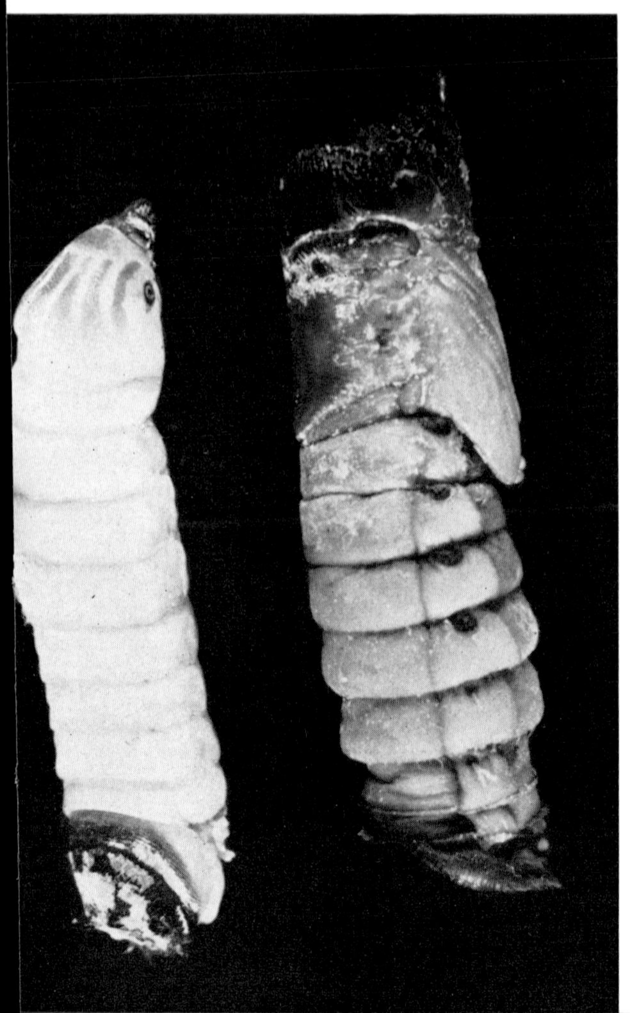

6 *Adult* females of the parasitic fly *Ascodipteron* (family Streblidae) embedded in the skin of the wing of a bat.

7 Larva and pupa of a pantophthalmid fly, a wood-boring larva from tropical South America.

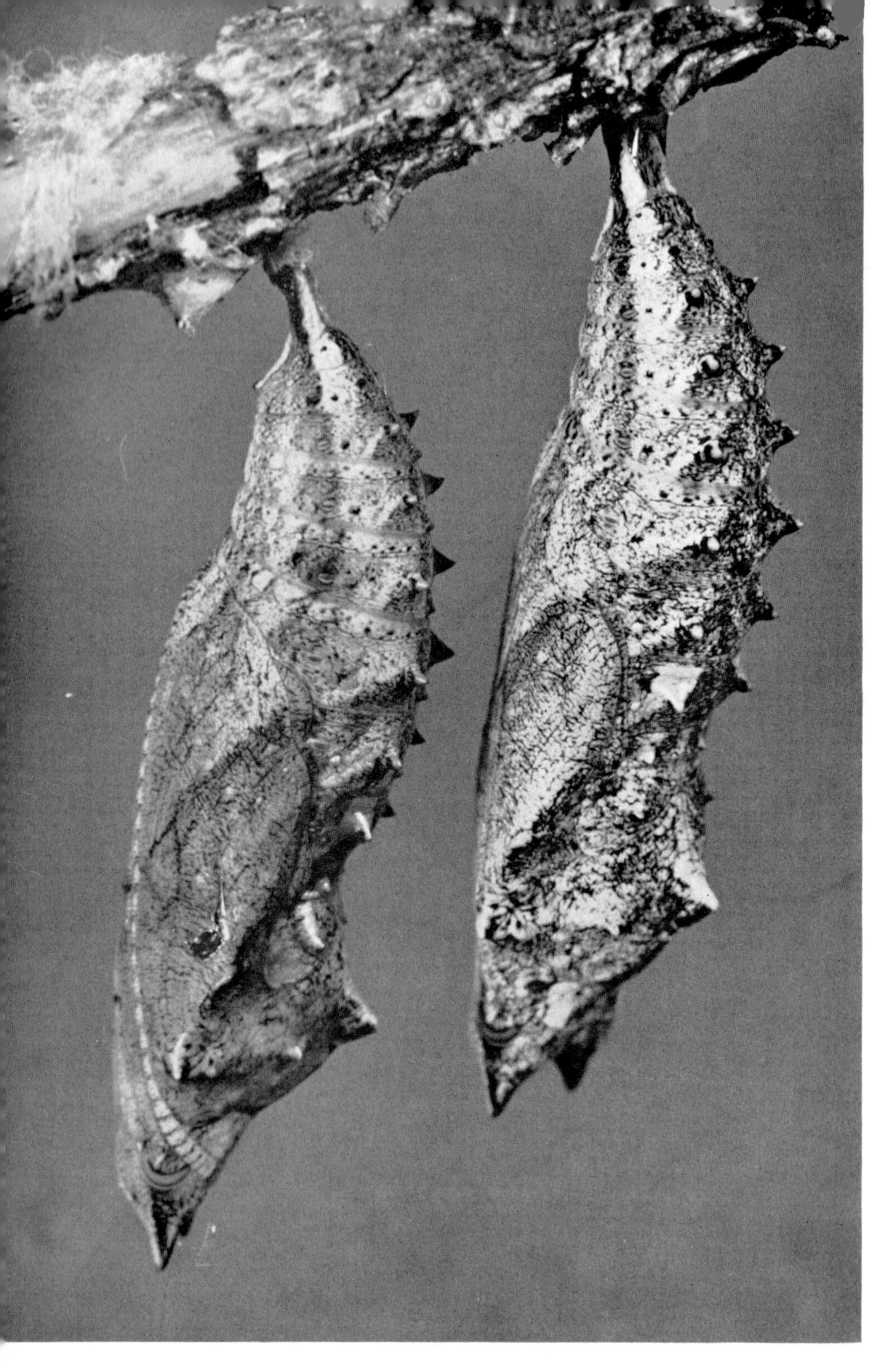

8 Pupae of the butterfly *Vanessa urticae* hanging from a twig. This pupa is *obtect*—appendages not freely movable—and *adecticus*—has no active mandibles which it can use to break out of a cocoon.

9 A pupa of a moth (also obtect and adecticus) inside its cocoon, a structure formed from extraneous materials.

10 Pupa of the Orange Tip butterfly, supported by a fine girdle of silk. Note how the wing-veins and the abdominal segments can still be seen, even though the external appearance has become stylised into a shape unlike that of any normal insect.

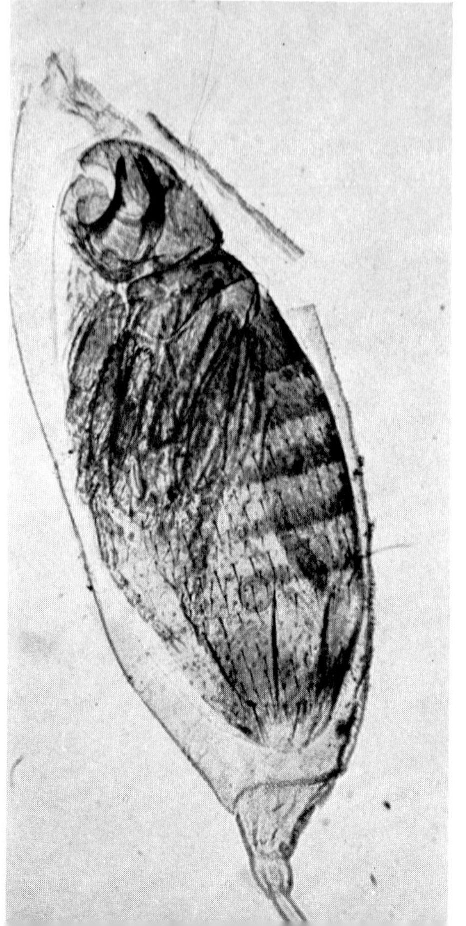

11 The pupa of this asilid fly is equipped with powerful spines, and works its way out of the wood in which the larva has been living. Note the T-shaped slit from which the adult has emerged; this is characteristic of the sub-order Ortho-rhapha.

12 Pupa of a fly of the family Phoridae, cleared and mounted on a slide to show the adult fly inside it. Note the respiratory 'horns' of the pupa (*top right*), which arise from the thorax.

13 A cockroach (Dictyoptera), one of the most ancient of winged insects.

14 A mole-cricket (Orthoptera) digs with its highly adapted fore-legs.

15 A short-horned grasshopper (Orthoptera) shows clearly its short antennae, mandibulate, biting mouthparts, and powerful jumping hind-legs.

16 An African bush-cricket (Orthoptera: Tetrigidae), highly modified for concealment among the vegetation.

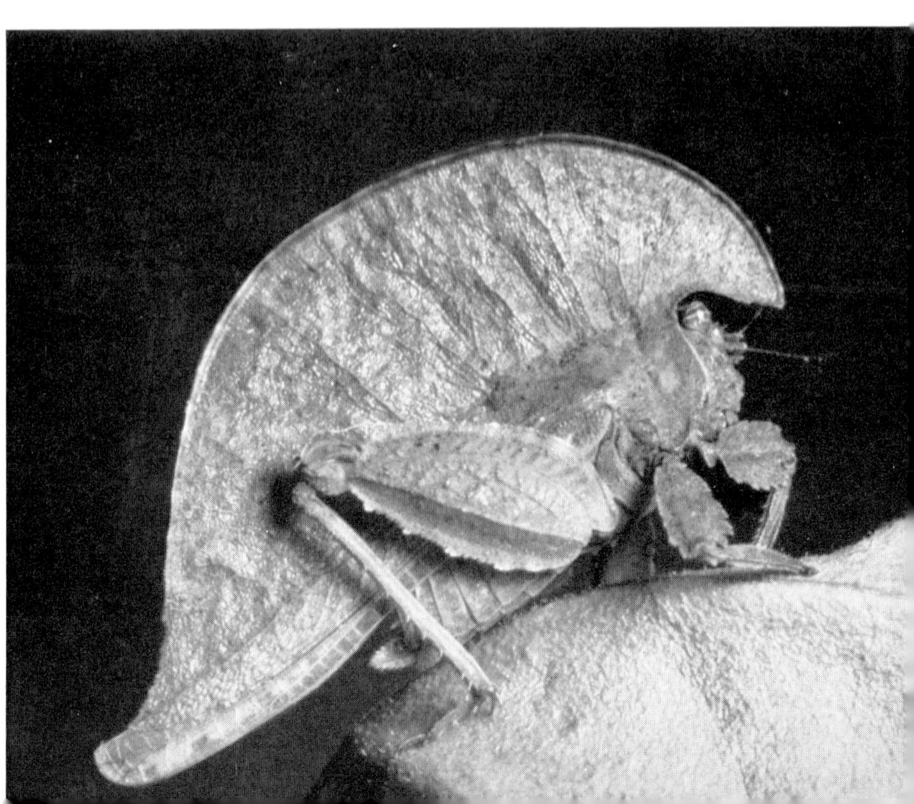

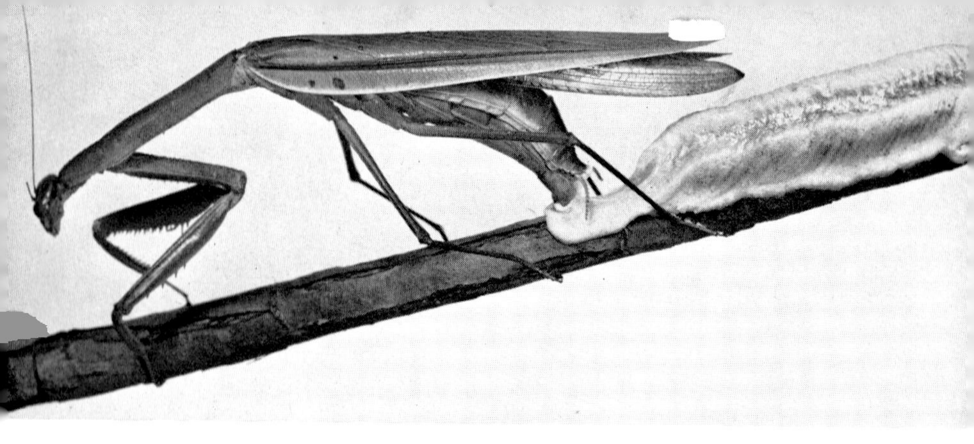

17 A mantid (Dictyoptera) laying eggs in a frothy egg-case.

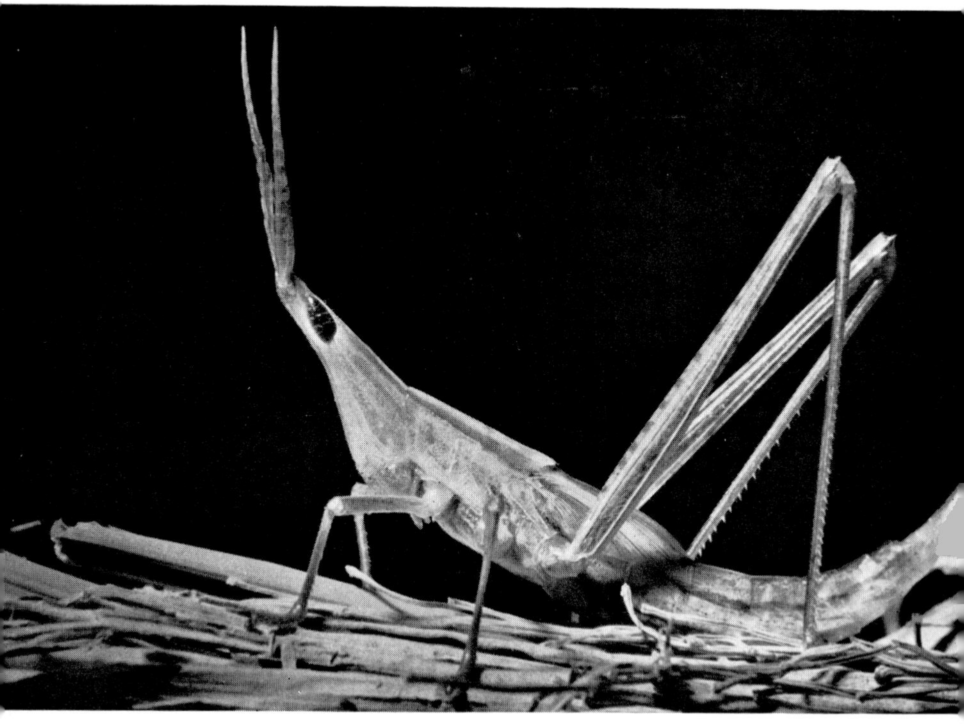

18 (*above*) A female slant-faced locust, with exceptionally long and thin hind-legs, as well as highly modified head.

19 (*left*) An earwig (mounted on a slide) showing the posterior cerci (forceps) and the fan-like wings (cf. fig. 25).

20 A dragonfly (Odonate: Anisoptera) newly emerged from its nymphal skin (*lower, right*).

21 A giant water-bug, a highly predacious insect, showing rapacious fore-legs, and mouthparts modified into a piercing rostrum.

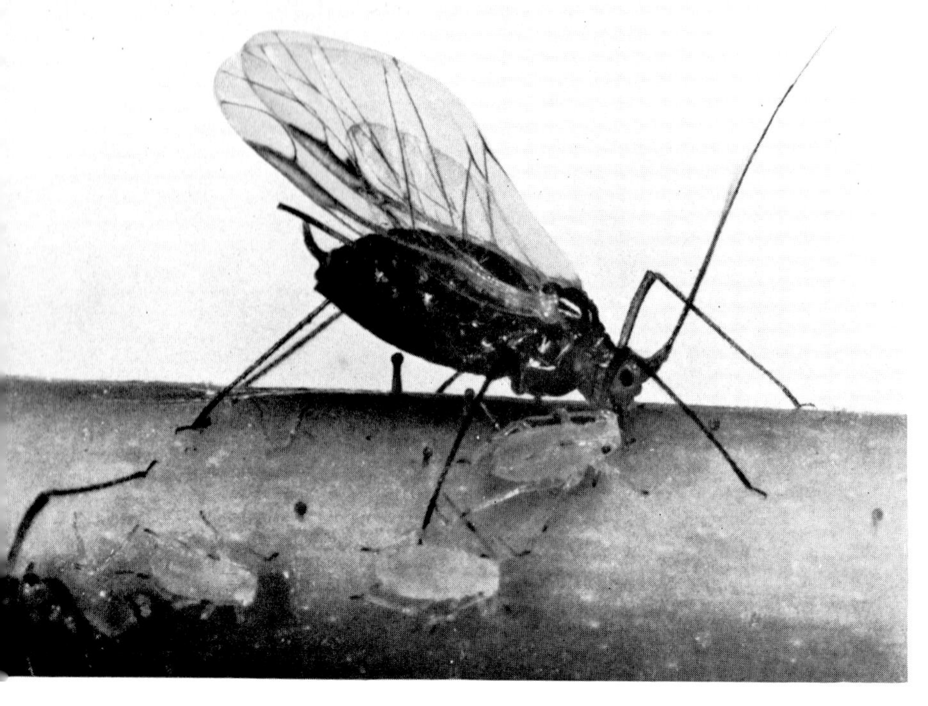

22 A winged aphid and nymphs.

23 Larva of a ladybird (Coleoptera: Coccinellidae) seizing an aphid. An example of an actively mobile *campodeiform* larva.

24 A lacewing fly (Neuroptera).

25 A newly hatched swallow-tail butterfly. Note coiled sucking proboscis and overlapping scales of wings.

26 An ichneumon-fly (Hymenoptera).

27 Head and thorax of a honey bee. Note fine, silky hairs; three ocelli at top of head; and pollen-carrying *corbiculae* on hind-legs.

28 The cockchafer beetle, *Melolontha*. Note *lamellate* antennae, which give an enormously increased area for the siting of sense-organs.

29 A rhinoceros beetle, with grotesque and seemingly useless elaboration of head and thorax.

30 A weevil, with mouthparts and antennae borne on an elongate snout. Note small size and position of eyes, indicating a way of life that is not dependent on acuity of vision.

31 Head and antennae of a male moth, elaborately developed sense-organs.

32 A ladybird in flight. Note the relatively passive role of the fore wings (elytra) compared with the active hind-wings.

33 A flying locust uses both fore- and hind-wings, which in this picture are both at the limit of their downstroke.

34 A luna moth flying, with antennae extended.

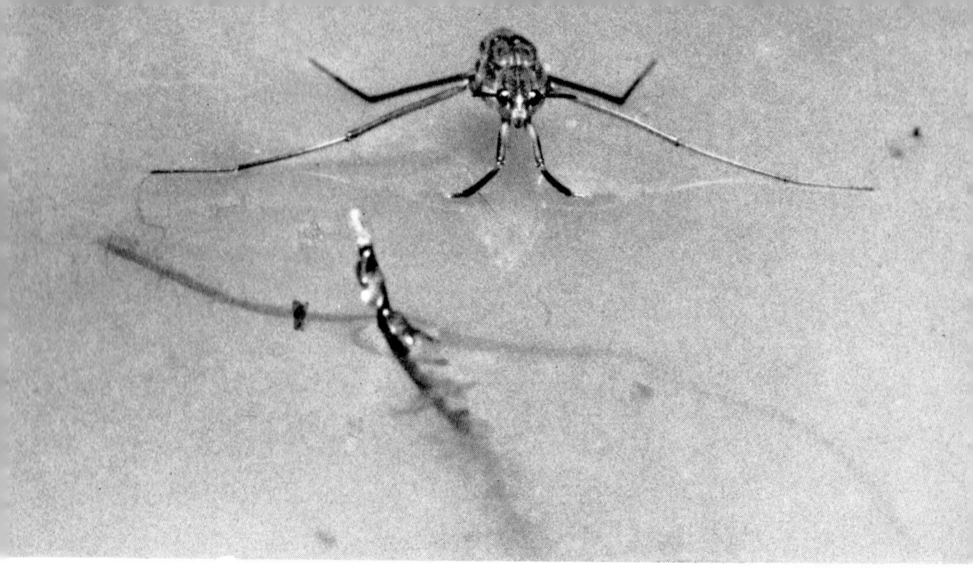

35 A water-strider (Hymenoptera: Heteroptera) supported on the surface film.

36 A carnivorous water-beetle attacks a leech.

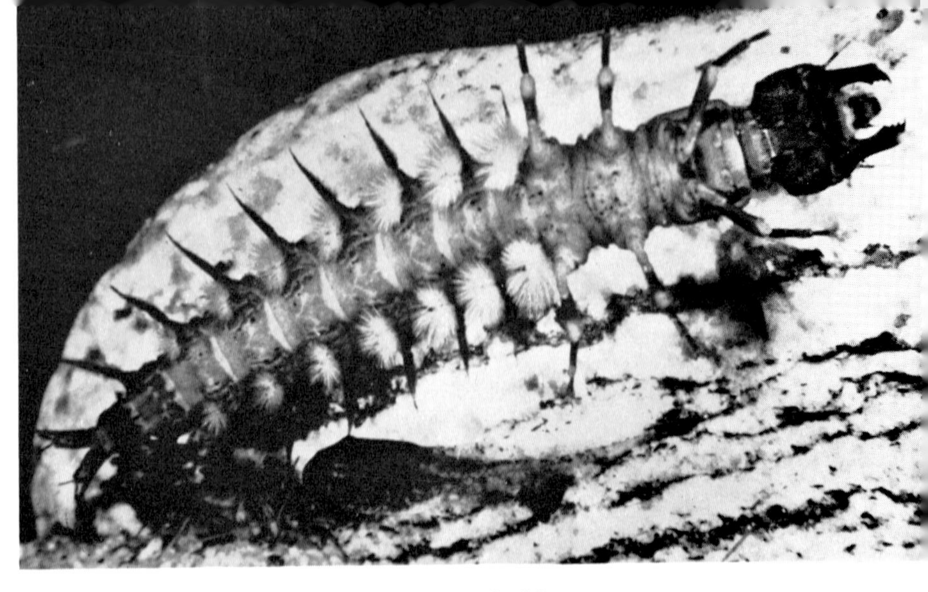

37 The carnivorous, aquatic larva of a Dobson-fly (Neuroptera: Megaloptera). Note the tufted tracheal gills on the abdomen.

38 Nymph of an assassin bug (Hemiptera: Heteroptera) attacks the larva of a fly.

39 Head of the monarch butterfly, showing coiled proboscis, formed from the two galeae of the maxillae.

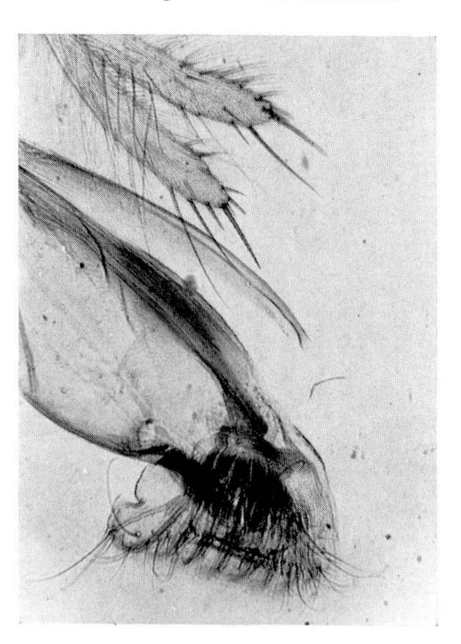

40 Mouthparts of a predacious dung-fly showing the pair of palpi (*above*), the labrum, and the labium, with its tip modified into *labella*, with tubular or groove-like *pseudotracheae*.

41 A 'supplementary queen' termite.

42 Aphids, adults and nymphs, feeding on a stem.

43 Four wood-ants dismembering a centipede.

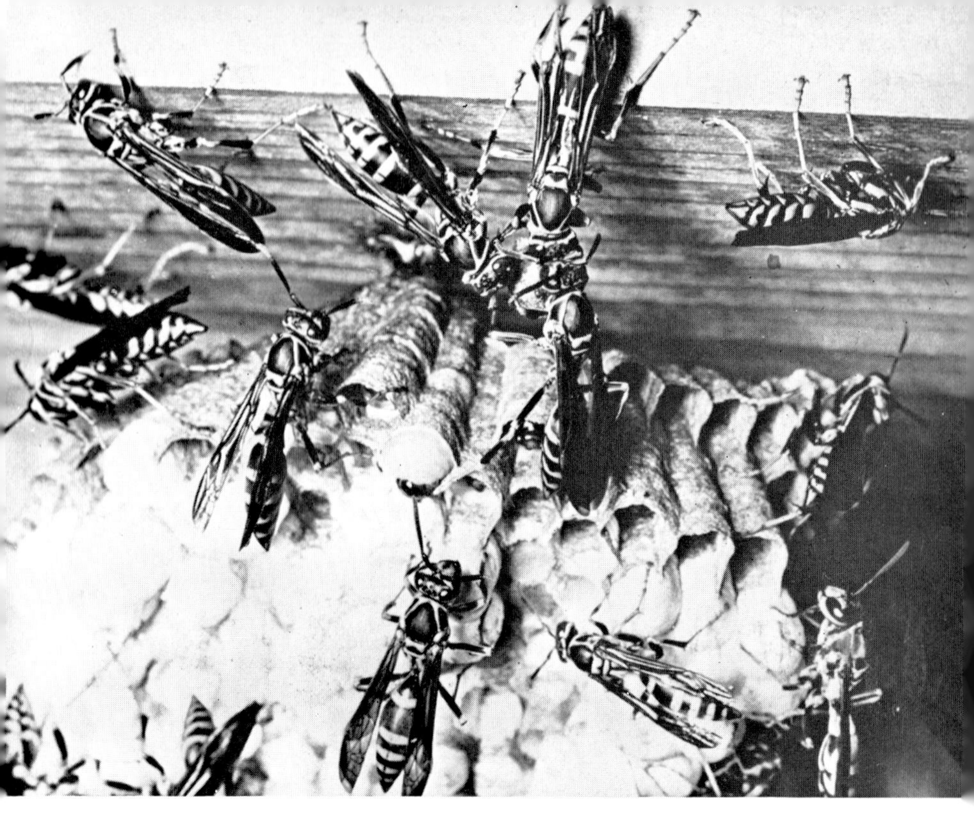

44 Wasps dividing food brought in by a forager; in centre a cell being capped over.

45 Hive-bees on a comb, showing larvae in open cells.

46 Ant carrying a pine-needle much longer than itself.

47 Trophallaxis, the exchange of food between members of a colony of social insects.

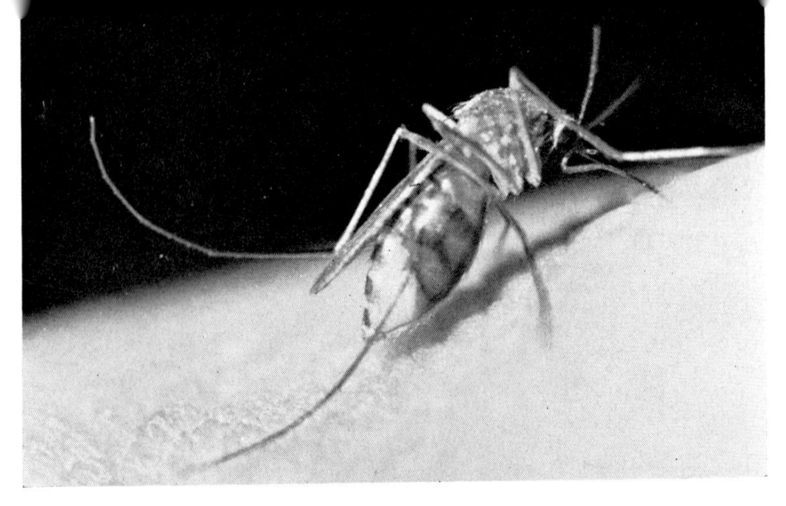

48 A culicine mosquito feeding from the human skin.

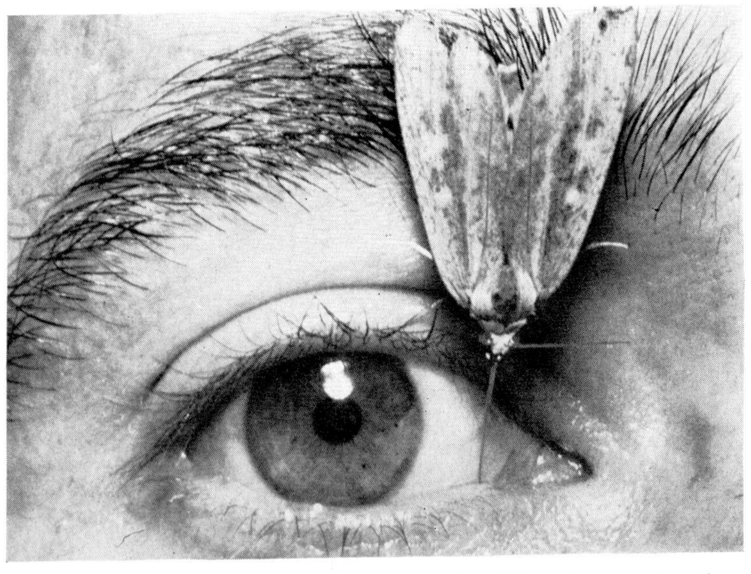

49 and 50 Two pictures of moths feeding from the secretion of the human eye.

(figure 62), a folding structure that can be projected at the victim of attack.

Carnivores among Hemimetabolous insects

No other order of mandibulate insects has gone over completely to a carnivorous diet. Mallophaga are often spoken of as 'biting lice', in contrast to the 'sucking lice' of the order Siphunculata, yet Mallophaga do not really bite. They are properly scavengers, chewing the debris of dead cells from skin, hairs or feathers. Though they are often referred to as parasites they do not really feed on the living tissues of the host. By pulling away at loose and flaking areas of skin they may set up irritation, and increase the risk of infection, just as we may if we scratch a sore on our own body.

The grasshoppers and their allies are pre-eminently plant-feeders, but some long-horned grasshoppers seize other insects, and of course the mantids are highly predaceous, even cannibalistic, feeding on their own species and even on their own mate. Among the related Dermaptera, some earwigs take living insects as prey; the wingless Hemimeridae live as apparent parasites of the African Giant Rat, *Cricetomys*, though probably they are only scavengers in the manner of the biting lice; and *Arixenia* lives in the fur of Indian bats.

It is probable that most scavenging insects occasionally take living organisms, including small insects. The question that concerns them is whether they can obtain food of a kind that they can masticate, without caring very much whether it is animal or vegetable. The gradual transition from feeding on dead animal matter to killing living animals is seen most clearly among aquatic insects. We have seen that many aquatic insects feed on detritus, helped by some kind of filter mechanism. There is always a good supply of dead animal matter – collectively known as *carrion* – in the water, and not content with that, a great many, perhaps most aquatic insects are carnivorous. In a similar way a high proportion of fish are carnivorous, suggesting that the aquatic life encourages carnivorous habits. It may be that prey is more easily seen in water, or that it is easier to filter or otherwise pick it out from water than from a mass of debris on land.

Though the first winged insects were terrestrial, they soon began to return to the water from which their ancestors came, at least during their immature stages, emerging into the air for their final, winged adult stage. Thus both ancient orders of Paleoptera, Ephemeroptera and Odonata, have aquatic nymphs. The nymphs of May-flies are mainly if not entirely herbivorous, and browse on algae and on underwater vegetation. By this means they get enough food, not only for themselves, but also to support the adult throughout its brief life, so that it does not need to feed. The nymphs of Dragon-flies, on the other hand, are carnivorous. Usually they take prey smaller than themselves, but some of the bigger ones will tackle worms, tadpoles and even small fish. It might be thought that such a diet would suffice for the whole life of the individual, yet adult Dragon-flies live much longer than adult May-flies, and feed voraciously by catching and killing other flying insects.

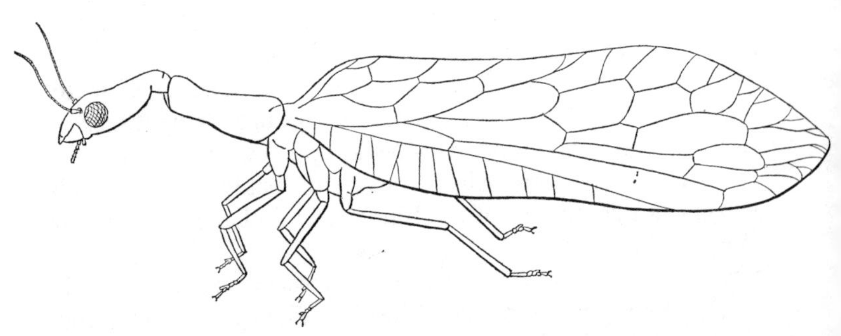

Figure 63. A snake-fly (order Neuroptera-Rhaphidioidea) with characteristic elongation of the prothorax (after Borror & Delong).

Stone-flies of the order Plecoptera are placed at the beginning of the Neoptera, the majority group of winged insects, but they are an archaic group, with several features in common with the May-flies and Dragon-flies. They also have aquatic nymphs, with a fully enclosed tracheal system. They compromise in their feeding habits. Some nymphs are carnivorous and prey on other small aquatic animals, while some feed on algae, or are general feeders. Some adult Stone-flies are said to feed on algae, and others not to feed at all, but in fact little is known about the ways of the adult insects, since they are retiring and furtive in their habits.

These three orders of insects illustrate the continuance of what might be called the primitive aquatic habit, a habit persisting from an early stage in insect evolution. The other hemimetabolous in-

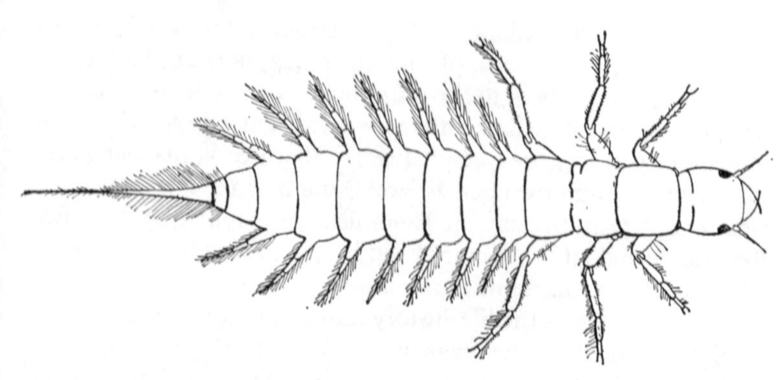

Figure 64. Larva of *Sialis* (Neuroptera: Megaloptera), with tracheal gills on the abdomen (after Ross).

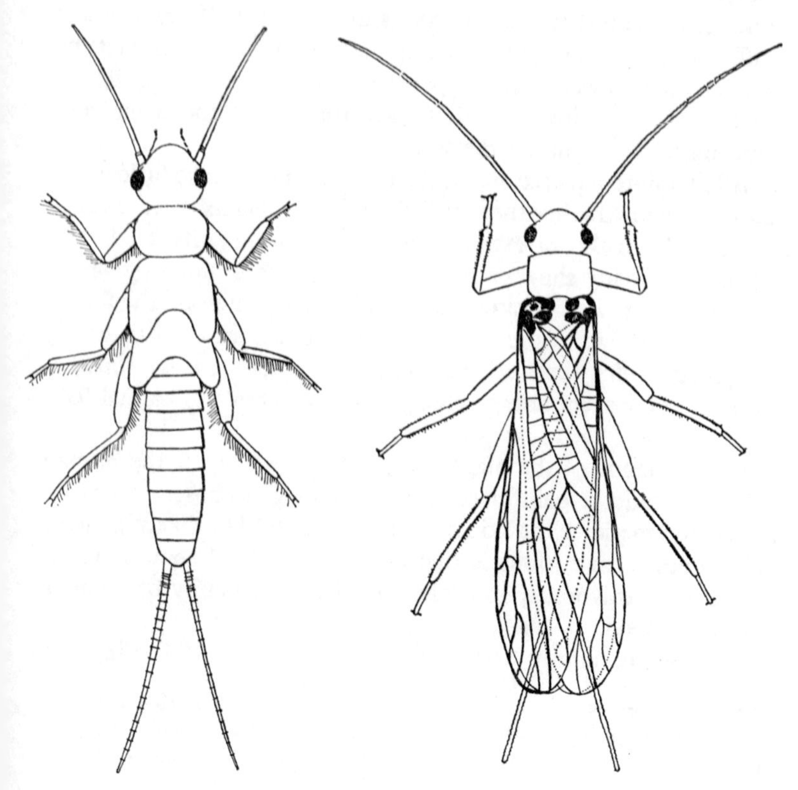

Figure 65. A stone-fly (order Plecoptera), nymph (*left*) and adult (*right*). Note visible wing-pads in nymph (after Ross).

sects were and are predominantly terrestrial, and those which have taken to water – as for example the water-bugs of the order Hemiptera – have moved back to the water much later in their evolutionary history. An immediate difference is that among the primitively aquatic insects it is the nymphal stages that live in the water; the adults emerge into the open air and live a free aerial life. There is thus among primitive aquatic Hemimetabola a foretaste of the two lives (larva and adult) that are lived by the more advanced insects which have a complete metamorphosis.*

Such a division of the life-history into two is not apparent in the secondarily aquatic water-bugs, where adults go into the water and feed along with their nymphs. Thus they are quite unlike adult Dragon-flies or May-flies, which cannot enter the water – though adult Dragon-flies sometimes descend to snatch insects out of the water – and would be helplessly trapped if they did. Adult water-bugs pass readily from air to water and back again. They overcome difficulties in respiration by taking down a film of air. In fact they are terrestrial insects, with open spiracles, living for preference in water. In feeding, too, they have transferred the habits of a land-insect to life under the water.

All Hemiptera pierce and suck, and it seems to make little difference to them whether they suck the juices of plants or of animals. Among the truly terrestrial bugs the great majority feed from plants, but many show carnivorous tendencies. Among the shield bugs of the family Pentatomidae, familiar large bugs on the foliage of the hedgerows, there are some, particularly in the sub-family Asopinae, which feed on the larvae of Lepidoptera, or those of beetles. Some seem able to change from vegetable to animal food indifferently.

Indeed, many of the bigger Heteroptera can pierce the human skin and inflict a painful bite. The piercing proboscis needs no adaptation to pierce the skin of an animal instead of that of a plant. What does seem curious is that some members of some families should be carnivorous, and others of the same family be content with plant food.

Thus in the family Reduviidae all three kinds of feeding are

*Some authorities emphasise this point by restricting the term Hemimetabola (=half-metamorphosis) to these three orders, and calling all those others in which the nymphs and adults live together the *Paurometabola* (=gradual metamorphosis).

known, vegetable feeding, sucking the body-fluids of other insects, and sucking the blood of vertebrates, including man. *Reduvius personatus* is a smallish, inconspicuous bug, with narrow body and long antennae, which preys on other insects, and is found in houses. If it gets the chance it will give a painful bite to the human skin. In South and Central America bugs of the genus *Triatoma* carry *Trypanosoma cruzi*, the flagellate Protozoon in the human blood that causes Chagas disease. The related Reduviid *Rhodnius* is a bloodsucking insect that has been used very extensively by Wigglesworth and his students in the investigation of many problems of insect physiology, partly because it is easy to feed, and to breed in captivity.

Carnivores among holometabolous insects

One characteristic feature of holometabolous insects is that the larva and the adult live independent and usually quite different lives, often in totally different surroundings. In general the larval stage is preoccupied with feeding, and the adult with reproduction. Often, as in Ephemeroptera, it is possible for an insect to accumulate enough food during its immature stages to make it unnecessary for the adult to feed at all. May-flies accomplish this even on a vegetarian diet, but they need a long time in which to do it, and may live from one to three years as nymphs before the very brief adult life. In contrast, the Dragon-flies continue to need a carnivorous diet as adults, even though the nymphs are also carnivorous, and long-lived as well. We do not know why these insects need so much protein.

Similar contrasts are to be found among the more advanced insects with complete metamorphosis, and the distribution of feeding between larva and adult is arranged to suit the habits of each group of insects. Either larva or adult, or both, may take up a carnivorous diet, making use of such mouthparts as it possesses, whether these are of a chewing type or are developed for piercing and sucking.

The order Neuroptera, placed first among holometabolous insects, gives a varied introduction to carnivorous habits. All are mandibulate, as befits a primitive group, and those which suck instead of chewing do so by means of grooved mandibles.

The Megaloptera have the chewing larvae and the Planipennia the sucking larvae. Those of *Sialis*, the Alder-fly, and *Corydalis*, the Dobson-fly, hatch from the egg on vegetation near water, which they enter, and there feed on insect larvae and other small creatures. They have typical biting mouthparts, with curved mandibles that are sharply toothed and very strong. The adults of these insects also have powerful mandibles, though it is thought that they probably do not feed. In the male *Corydalis* the mandibles are exceedingly long and sickle-shaped (figure 37).

The larvae of the Snake-flies, *Raphidia*, are terrestrial and run about under loose bark, but they have similarly curved mandibles, with which they catch and eat terrestrial insects just as voraciously as their aquatic relatives. Note how similar is this larva to the larvae of the primitive ground-beetles, and even to the adult of certain families of beetles, such as Staphylinidae (rove beetles). These particular beetles, like the *Raphidia*, are carnivorous and chase their prey by actively running after it on their well-developed thoracic legs.

Among the Neuroptera of the sub-order Plannipennia there is even more diversity of habit, though the larvae are all predaceous. They also have curved, sickle-shaped mandibles, but the maxillae have the same shape, and each fits closely to its corresponding mandible to form a sucking tube. Between the two paired organs the insect is pierced and sucked dry.

The food of the larvae of the Lacewings – Green (Chrysopidae) and Brown (Hemerobiidae) – is chiefly aphids and other soft-bodied Homoptera, along with mites. The larvae roam about eating these in enormous numbers as they encrust the stems of plants. Once again they show an interesting parallel evolution in shape and even in pattern to beetles of similar habits: in this case especially to the larvae of ladybirds (Coccinellidae) (plate 73).

The larvae of Nemopteridae, Ascalaphidae and Myrmeleontidae are of a somewhat different shape; broader, less streamlined, with powerful mandibles in a large, mobile head. The appearance of such larvae suggests that they are poised ready to twist quickly in any direction, and to pounce on their prey. This is exactly what they do. Instead of browsing on sedentary insects such as aphids, which offer a sitting target, these larvae find more active prey in dry, dusty situations. The Nemopterid larva lies concealed in dust, and pounces on Psocids and other small insects. Some larvae of Myrmeleontidae do the same, or catch ants on tree-trunks – a sporting way of hunting, if ever there was one – but the most familiar larvae, from which the family gets its scientific name are the 'ant-lions' (from *Myrmeleo* and *Leo*). These dig conical pits in fine dust and lurk there, feeding on ants and other insects that are unlucky enough to fall in.

At the other extreme there are Planipennia with aquatic larvae. Those of Osmylidae live under stones and prey on other aquatic larvae, especially those of Diptera; larvae of Sisyridae cling to fresh-water sponges and suck their tissues. Both families have the mandibles and maxillae only slightly curved, more like the sucking mouthparts of other orders.

Adult Plannipennia, if they feed at all, are also predaceous. This is carried to an extreme in the family Mantispidae, the adults of which look remarkably like a miniature praying mantis as in 24. There is the same elongation of the prothorax, and the same elaboration of the fore-legs into a cruel pair of nutcracker-like organs with which the prey is seized. Larval Mantispids enter the cocoons of Lycosid spiders and feed upon their young. The first-stage larva is campodeiform, and actively searches for its food; later larval instars are soft-bodied and swollen (scarabeiform). This is another example of *hypermetamorphosis*, and illustrates how this form of adaptation can arise sporadically if the insect lays eggs in a place remote from the food-supply of the older larvae.

The order Neuroptera has been discussed at some length because they provide such a varied selection of examples of carnivorous feeding in both larvae and adults. The other smaller orders of Holometabola are each more uniform in their habits, but naturally some of the big orders have evolved within themselves a variety of feeding habits.

Mecoptera, the Scorpion-flies, have mandibulate mouthparts of a chewing type, both as adults and as larvae. Though the larvae look like caterpillars they feed on animal matter, and are considered to be mainly scavengers, though they may take some living insects. This is also true of the adults' feeding habits, though they look aggressive enough with their mouthparts at the end of a 'beak' (figure 38). Those of the family Panorpidae increase the appearance of ferociousness with a curved tip to the abdomen, from which they get their name of 'Scorpion-fly', though this is not a sting!

The Caddis-flies of the order Trichoptera are also mandibulate, both as adults and as larvae. Among the adults the mandibles are often reduced or lost, and the maxillae particularly adapted for licking and sucking up liquids in a manner which foreshadows the sucking proboscis of Lepidoptera. Nearly all caddis larvae are aquatic, and most are detritus-feeders. Those which live in cases of silk covered with small spores maintain a current of water through the case by undulating movements of the body. Certain carnivorous caddis larvae spin nets of silk with which they catch their prey in the water (figure 39).

Lepidoptera, one of the biggest orders of insects, is remarkable for the fact that no adults and only a very few larvae are carnivorous. The adults, as we have seen, have lost the mandibles almost completely and feed, if at all, by means of a proboscis developed from the maxillae only. This can take in only liquid foods from exposed surfaces, and is not rigid enough to pierce the skin of vertebrates, or even of other insects.

Larvae of Lepidoptera, the familiar caterpillars, are plant-feeders *par excellence*, yet, as among other mandibulate insects, the tendency to change to a carnivorous diet is always present. Insects that feed superficially on plants are particularly likely to seize and eat aphids, coccids and other sedentary Homoptera, which must be a tempting morsel. In the family Blastobasidae the Acorn Moth, *Valentinia glandulella*, feeds internally in acorns, while in the same family, the larvae of *Zenodochium coccinivorella* feed upon scale insects of the genus *Kermes*, the females of which are sedentary and look like plant-galls. In the Australasian region several small families of moths – e.g. Cyclostomidae and Epipyropidae – also have larvae that are parasitic on Homoptera.

The moths mentioned above are in the superfamily Tineoidea, relatives of the Clothes-moth, *Tinea*. Carnivorous habits break out again in a rather unlikely place among the 'Blues', butterflies of the family Lycaenidae: the larvae of a number of species feed on aphids and coccids, or even enter ants' nests and eat the larvae and pupae there. They are not always attacked by the ants, which are easily pacified with a few drops of sugary liquid (see *trophallaxis*, Chapter 17). Some Lycaenid larvae have special abdominal glands from which they produce a liquid that pacifies the ants, and protects the carnivorous caterpillars against attack.

Among Pyralid moths, many of which feed on grasses, grains, seeds and generally rather dry, farinaceous foodstuffs, there is a coccid-feeding species *Laetilia coccidivora*. Again, the transition has probably been easy to make during evolution.

In contrast to the mainly herbivorous Lepidoptera the five remaining orders of Holometabola have many carnivorous members. Taking first the *Diptera*, the true flies, the sucking proboscis of the adult is of a type that can readily be adapted to pierce the skin of other insects, or of vertebrate animals. Indeed it seems likely that the ancestral flies had this ability, and that the predaceous and bloodsucking flies of today fall into two groups: those that have retained the ancestral habit, and those that have lost the ancestral habit, but have subsequently become carnivorous again by an independent evolution, of a different mechanism.

Flies of the first group use the mandibles and maxillae as piercing stylets in the same way as these are used by Hemiptera (figure 53). Such are mosquitoes, Black-flies, biting midges, Horse-flies and a few others. Most of these take blood from vertebrates, including man; some of the biting midges suck the blood of other insects – e.g. *Forcipomyia* attaching itself to the wings of Dragon-flies and sucking from the veins. These carnivorous flies all belong to the more primitive families. Mostly they have larvae that are aquatic or semi-aquatic, which may also be carnivorous as among mosquitoes and Horse-flies. They are paralleled by a group of non-biting midges and other flies with mainly terrestrial larvae, and from this second stock has apparently evolved the rest of the Diptera. Along this line the mandibles and maxillae have been lost as piercing organs, and the principal sucking equipment is the sponge-like labella, with the pseudotracheal tubes. The proboscis of the House-fly (*Musca*) is an example of this kind. Yet along this line of evolution which leads to the most flourishing and evolutionarily successful flies of today, there have been many independent excursions into carnivorous habits.

Thus three whole families have become fundamentally predaceous on other insects: Asilidae, the Robber-flies; Empididae, the Dance-flies; and Dolichopodidae, which have no common name. The first two pierce and suck in the old way, but having once reduced the mandible and maxillae they have to rely mainly on the sharp hypopharynx, with its salivary duct, piercing the prey as with a hypodermic needle (plate 40). The third family Dolichopodidae, are really committed to a spongy proboscis of the House-fly type, but they have ingeniously made use of it to envelope and squeeze small soft-bodied insects. With the aid of specially hardened pieces of chitin they burst the integument of their prey and suck it dry before discarding the empty skin. A similar habit has been developed independently in other families, notably in some Ephydridae.

Among the 'higher' flies, that is those that have evolved away from the primitive habits, there has been sporadic reversion into blood-sucking in a few families. In the family Muscidae, which includes the House-fly, there are many species that settle on grazing animals – or on man – to suck the sweat and other fluids from the body-surface. Many have acquired a taste for the blood that oozes from wounds, or from the punctures made by bloodsucking insects such as Horse-flies. Non-biting flies often gather in swarms around the biting flies, and may even drive them away to get at the flow of blood. Several species – e.g. *Musca crassirostis*, a close relative of the House-fly – assist the flow of blood by scratching at the skin with hardened 'teeth' similar to those we have just mentioned in Dolichopodidae.

Finally among flies, some of the muscoid group have reverted completely to sucking blood, and have contracted the spongy labella into sharp piercing organs, and stiffened the stem of the labium until it can be pushed bodily into the skin of the victim. Such are the Tsetse-flies, *Glossina*, their relatives the Stable-fly, *Stomoxys*, and one or two more, as well as the three families of Pupipara which feed on birds and on mammals including bats. It is remarkable that these bloodsucking flies of later evolution should have the same diet in both sexes, whereas among mosquitoes and other primitive bloodsuckers only the females suck blood.

The more advanced flies of this group spend their larval life as a maggot or grub. A maggot is pointed in front and blunt posteriorly, whereas a grub is more fleshy and barrel-shaped, but their essential structure is similar. There is no head-capsule, and the mouthparts take the form of *mouth-hooks* (figure 72c). There is dispute as to whether these are modified from the original mandibles and maxillae, or whether they are structures specially evolved. Though deceptively simple in appearance, they have proved immensely versatile organs for feeding. Most often they are used to feed on soft, decaying materials of all kinds, whether vegetable or animal in nature. The mouth-hooks are also hard and sharp, and deal effectively with the firm tissues of either plants or animals. Some of these larvae burrow deeply into the stems of plants, or make mines in the leaves; others equally effectively enter the bodies of vertebrates by way of wounds, or natural orifices, or by piercing the skin. They may feed on natural fluids, or exudations from wounds, or directly on the living tissues.

There is thus a natural progression among fly-larvae from vegetable material to animal carrion, thence to becoming a parasite on the living tissues of a vertebrate animal or of another insect. Sometimes one and the same species may be capable of being a carrion-feeder or a parasite, according to opportunity. A well-known example is the Sheep Maggot-fly, *Lucilia sericata*. The larvae of this fly are common in dustbins, butchers' shops, abattoirs, etc. When they have access to sheep they often enter the soiled wool round the anus and the crutch, and besides feeding on organic debris there they irritate and finally penetrate the skin of the sheep, making a suppurating sore in which they can feed to repletion.

Those larvae that are referred to as 'grubs' are usually those that have become *obligatory parasites*, that is have ceased to feed externally, and now spend their whole growing period within the tissues of the host-animal. For this the basic equipment of such a larva is excellent. The mouth-hooks serve to grip and lacerate the tissues, and are often powerfully curved to this end: the mouth-hooks of the Oestrid larva are usually more convex than those of the House-fly. Maggots and grubs typically have two pairs of spiracles, one on the thorax and a bigger pair at the extreme tip of the body. The posterior spiracles enable the larva to burrow deep into the tissues of the host, and still communicate with the outside air. Such grub-like larvae as that of the African Tumbu Fly, *Cordylobia anthropophaga* live and feed in a boil-like swelling on the skin of their unfortunate host, which is often man.

Fleas are so different in structure from any other insects that they are given a separate order Siphonaptera, a name expressing the fact that they are wingless sucking insects. Most classifications place them near to Diptera because their larvae are somewhat similar. The larvae do not suck blood, but live in the debris round the nest or habitation of the host animal. The larvae are believed to feed on organic debris from the skin of the host, as well as on scraps of dried blood passed *per anum* by the adult fleas, so that their diet is essentially a carnivorous one. Adult fleas, of course, are entirely blood-sucking. They have a piercing type of proboscis constructed from the mandibles and maxillae, plus the labium and epipharynx, all blade-like. Thus all members of this order of insects feed exclusively on material of animal origin.

Adult Hymenoptera are usually equipped with mandibles, even if they have modified other mouthparts into a sucking apparatus. The Saw-flies (Symphyta) are essentially vegetarian, but a great many of the waisted Hymenoptera (Apocrita) take animal food either as adults or as larvae. Some do both; an example is the parasitic wasp, *Habrocytus*, which does so, first as it pierces the cocoon of its host, and then as it sucks at the hole thus made, and so shares the food of its own offspring.

When it comes to larval feeding, very large numbers of parasitic Hymenoptera are predominantly flesh-feeders, living internally in the eggs or larvae of other insects. The larvae of these Hymenoptera are white, legless grubs, rather like some larvae of flies, but having arrived at this structure by quite different evolution. The maggot of the higher flies appears to be an example of *pre-adaptation* – that is the larva arose in evolution a long time ago, and has proved so versatile that flies that have this larva have been able to take up a great variety of different ways of life. The legless larva of Hymenoptera, in contrast, seems to have evolved in direct response to the sheltered life of a parasite within the tissues of its host.

These conditions apply only if the adult insect lays its eggs on or in the food-material, so that the larva can live a sedentary existence from the very beginning. Among many parasitic Hymenoptera the larva has to find its own food, and in these genera the first stage larva is a more active form. It still has no true thoracic legs; having once lost these in evolution, it does not acquire them again, but instead it has evolved various secondary extensions of the body surface which help it to move actively during the short time available to it during this stage of life.

The best known example of this is the *planidium* first-stage larva of the Chalcidoid genus *Perilampus* which has developed strong, spiny processes on the ventral surface of each segment, and pushes itself along with these. Other types of primary larva have been given names, but it is not necessary to enumerate them here. All life-histories of this type are examples of the phenomenon of *hypermetamorphosis*, by which the onus of finding a host is transferred from the parent insect to the young larva. The same device has been evolved independently in other orders of insects, including many parasitic Diptera. The later larvae of such insects, having arrived at their feeding-place, do not need to move about again, and are fleshy, immobile grubs.

Larvae of parasitic and aculeate Hymenoptera have mandibles of fairly simple shape, but evidently derived directly from the primitive mandibles of generalised insects, and not a special evolution as has been suggested for the mouth-hooks of cyclorrhaphous Diptera.

As among Diptera, carnivorous larvae may feed on carrion – dead animal tissues – or may transfer to living tissues and become ectoparasites or endoparasites of vertebrate animals or of other insects. Among Hymenoptera there are none that parasitise vertebrates – a rather remarkable limitation for such a versatile group. We ignore, of course the bites and stings that Hymenoptera may inflict upon us in a purely defensive spirit. Carnivorous Hymenoptera feed entirely on other Arthropods, mainly upon other insects.

Many wasps, for example, give animal food to their larvae at least in the later larval stages if not at first. The social wasps, which include the common garden wasps and the hornets, tend their young continually (see Chapter 16), and give them scraps of insects captured and killed by the adult wasps. The solitary wasps provision the cell with caterpillars or other insects which they have stung into immobility, but not always killed. Yet this, too is really carrion-feeding. The only value of paralysing the prey instead of killing it outright is that the processes of decay are thereby arrested, and when the egg hatches the larva is provided with fresh meat instead of that which is already rotting.

Wasps belonging to a few families are true parasites – e.g. Scoliidae which parasitise beetle-larvae, *Rhopalosoma*, which is an ectoparasite of a tree-cricket, and the well-known Mutillidae, vividly coloured wasps which are all parasites of other insects, including some solitary bees and wasps.

Thus as often happens with descriptive names used in classification, the Hymenoptera-Parasitica are not all parasites, and there are some parasites among the contrasting group of Hymenoptera-Aculeata. The two orders Diptera and Hymenoptera illustrate how difficult it is to draw a line between a carnivore and a parasite.

Ants are really only one family of insects, but perhaps the most varied and fascinating family of all. All ants are social insects (see Chapter 17) and their versatility lies not so much between one genus and another as among the various *castes* which go to make up one species, and which share life in the same nest. Many ants are carnivorous, including the notorious Driver Ants (Dorylinae) and Bull-Dog Ants (Ponerinae) of the tropics. In this vigorous group of insects one cannot draw a line between herbivores and carnivores, and it is enough to say that most ants will eat anything that is to hand, whether vegetable or animal.

Turning now to the last remaining big order of insects, the beetles of the order Coleoptera are again a group of insects equipped with mandibles both as larvae and as adults. Though many beetles fly readily and well, in fact their most characteristic activity is that of crawling about among debris on the ground, or over vegetation. It is probable that in origin beetles were omnivorous feeders like cockroaches. Quite early they must have moved, by way of carrion-feeding towards a carnivorous diet. The most primitive families among the beetles that exist today, are those of the sub-order Adephaga, which includes the Carabidae, the familiar black ground-beetles, the Cicindelidae, or Tiger-Beetles, and the Dystiscidae, the voracious water-beetles – and almost all are carnivorous. Moreover, the larvae are as fierce as the adults (plates 23, 36).

Among the rest of the order Coleoptera there are many carnivorous forms, but in general these occur sporadically along with close relatives which may eat decaying animal or vegetable materials, or scavenge in the nest of mammals, birds or other insects, picking up a mixed protein diet from the leavings of other animals.

Beetles therefore give the impression of an order that was mainly carnivorous when it first came into existence, but which in the course of a long and highly successful evolution has produced many newer groups that have turned back again to a vegetarian diet. These include most of the beetles that are best known to us, because they eat things which matter to man: they bore in wood, or into plants, attack the roots of growing plants or, among the most troublesome of all, feed among the grain and other foodstuffs that we store for our own use.

The parasitic *Strepsiptera*, which are often considered to be closely related to Coleoptera, might on this view be regarded as an early offshoot from a carnivorous ancestral stock.

15
Social insects 1: Termites

The term *social life* is used in entomology in a narrower sense than in ordinary life. It excludes all forms of gregariousness, which is the assembly of a number of individuals of the same species and at the same stage of development: e.g. the larvae of the Army Worm, or the hoppers or the adults of a swarm of locusts. The ways in which these mostly affect Man are briefly described in Chapter 22.

The essence of social life is not numbers, but interdependence, which originates as *maternal care*. The very large majority of female insects lay their eggs as an instinctive physiological act, and take no further interest in them. Of course most egg-laying females drop their eggs in a place where the eggs will have a good chance of hatching, and where the larvae will find food. Some, like the ichneumon fly *Rhyssa*, which penetrates with its ovipositor the bark and wood of trees to reach larvae of Saw-flies beneath, seem almost to perform a deliberate act of skill and cleverness. Seen from a human point of view, this feat arouses admiration at the cleverness and surprise that such a difficult feat should be necessary when other insects get by with just dropping their eggs to the ground.

Yet it is still true that even such skill is no more than an elaboration of a basic instinct, perfected by natural selection. The female goes through its ritual of egg-laying, but it never sees its eggs again any more than if it had merely dropped them in flight.

Even the provisioning of its cells by a solitary wasp is not true social behaviour, though it may lead towards it. Instead of bringing its eggs to the food wherever that may be, the wasp encloses both food and eggs in a cell, where the larva can feed in peace and security. But most solitary wasps do not take any further interest in their offspring. Having been provided with a store of food the larva is left to take its chance. When the wasp keeps its cell open and continues to bring food to the larva it has taken the essential step that leads to social life – the continuing contact between successive generations which is implied by the term *maternal care.*

Mention of social life brings to mind especially the Hymenoptera, wasps, bees and ants, which are among the most highly evolved of insects. It is tempting, therefore, to think of social life as a crowning achievement of insect evolution: as indeed it is in some ways. Yet the other spectacularly successful social insects, the termites, are a primitive order far removed from Hymenoptera in the evolutionary scale. Social insects are not, therefore, in one line of evolution that has been steadily perfected from primitive beginnings.

On the contrary, there are many examples of maternal care – and with it a rudimentary social behaviour – from the Dermaptera upwards. Most of them have never progressed far except, for some reason not known to us, the termites near one end of the evolutionary scale, and the Hymenoptera at the other. In spite of a superficial similarity, we shall see that the social organisation in these two orders is quite different, and obviously originated quite independently.

Rudimentary forms of maternal care

The behaviour of female of the common earwig is usually quoted as the basic example of rudimentary maternal care. She lays her eggs in a small hollow in the ground, and not only covers and protects

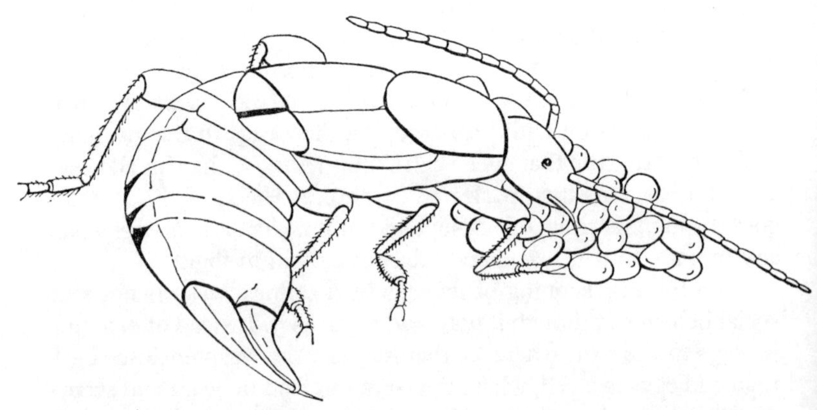

Figure 66. A earwig (order Dermaptera) 'brooding' its eggs, and thus showing rudiments of maternal care (after Ross).

them, but actively turns them over and licks them. It has been suggested that without this attention they become mouldy and die. She stays with the young nymphs for a few days after they hatch, and then deserts them. She does not obviously feed or otherwise tend the nymphs. The number of eggs thus guarded varies from about twenty to perhaps as many as ninety (figure 66.)

The evolution of this habit in earwigs is clearly an adaptation to meet the hazards of their particular oviposition site. Related primitive insects protect their eggs by enclosing them in an ootheca, a tough protective envelope – as in cockroaches and stick- and leaf-insects – or by pouring over them a secretion which hardens, as in locusts and mantids (plate 17). In this state these insects abandon their eggs, which fall a prey to many insect predators and parasites, and even to cannibalism if the early arrivals eat some of the remaining eggs. Maternal attention by the earwigs avoids most of these hazards, and perhaps enables the insect to lay fewer eggs and still survive.

It is often noticed that one group of insects appears to have evolved an ingenious device to do something which nearly all other insects contrive to do without nearly so much trouble. Maternal care of the eggs and young larvae is an example of this. Examples of it occur sporadically in other orders, but most often in the orders that are fairly highly evolved.

Among Hemimetabola there are 'brooding bugs', notably in the primitive family Pentatomidae, the Shield Bugs, often called Stink Bugs because they can emit an unpleasant odour from thoracic glands. Some Pentatomids remain with their eggs, and for a time with the newly hatched nymphs, though not giving them such attention as do the earwigs. It is interesting that this method of direct maternal protection has been evolved among the Heteroptera, whereas several families of the Homoptera protect their eggs and very young nymphs by covering them with froth or cuckoo-spit – as in Cercopidae, the Frog-Hoppers – or with white wax as in the Psyllidae, Aleurodidae and Aphididae.

Diptera, another highly evolved order, have several times developed the habit of protecting their eggs. The female of *Goniops chrysocoma*, a North American Horse-fly, lays her eggs on a leaf overhanging the water, attaches herself to the leaf by digging her claws into it, and dies in this position, her body acting as a protective cover for the eggs. The females of *Atherix ibis*, belonging to the related family Rhagionidae also brood to death over their egg-masses, assembling in large numbers for the purpose until the leaf is covered with what looks like a swarm of bees. It used to be thought that when the eggs hatched the larvae fell at once into the water, but now it is said that the first instar is passed under the protection of the massed females, and that the larvae do not fend for themselves until they are in the second instar.

The first instar larva of many insects is small and vulnerable. During artificial rearing this is often the most difficult stage to bring through successfully, and there is inevitably a high mortality. The very young larva, if it feeds at all, must obviously find food quite different in size as well as in composition from that which satisfies the bigger larva. One way of overcoming this difficulty is obviously to lay a large number of eggs and to accept a heavy loss; the method of the herring. Another is to provide the eggs with enough food to suffice for the first instar larva as well, and then to protect this by maternal attention, so that the larva does not have to live on its own until it is bigger and stronger.

This view of maternal care may be compared with another device, that of hatching the eggs internally and laying living larvae. Such insects are said to be viviparous, or ovo-viviparous. This is fairly common among parasitic insects such as Hippohoscidae among flies.

Though beetles often occur together in large numbers we should think of them as gregarious rather than as social insects. Yet there are several groups in which the males and females both look after the eggs and larvae, and come significantly nearer to a rudimentary social life.

Certain timber beetles of the family Scolytidae (sometimes divided into Ipidae and Platypodidae) live in tunnels in growing trees, feeding not upon the wood, but upon fungi which grow there, and which are carried from one breeding-site to another as spores on the bodies of the beetles. They are sometimes called 'ambrosia beetles' in reference to this fungus or 'ambrosia'. The adults regularly feed the fungus to their larvae, tending them until they are ready to pupate, and even removing the faeces of the larvae from the galleries. In some genera the adults not only live gregariously, but even polygamously, one male in association with a number of females. In *Xyleborus* up to sixty females are said to associate with one male.

Even this degree of social life is exceeded in beetles of the family Passalidae, which live in decaying wood. The male and female of *Passalus cornutus*, for example, chew the wood and feed it to their larvae, tend them throughout their larval life, and even help them to make cocoons. Both adults and larvae stridulate – i.e. make a noise by rubbing parts of the body together (see Chapter 9). The larvae have sacrificed the third pair of legs for this function, reducing them to little more than the enlarged trochanter, which is rubbed against the coxa of the middle leg. It is thought that the sound thus produced helps the community to keep together.

The larvae of the big stag-beetles of the allied family Lucanidae also make sounds. The larvae live in the decaying stumps of trees, but little is known about their habits, how close is the association of different individuals, and whether sound-production has any relation to social life.

The closely related family Scarabaeidae contains many well-known dung-beetles, including the famous Sacred Scarab of Ancient Egypt, and *Geotrupes*, the more commonplace dung-beetles of the temperate meadow-land. Larvae of *Geotrupes* stridulate with a reduced hind-leg as do the Passalidae. Several genera of dung-rolling beetles of the sub-family Coprinae go further. The male and female make a brood-chamber in a ball of dung, with cell-like subdivisions, in each of which is laid one egg. The female guards the brood, and may even survive the upbringing of one brood and lay a second batch of eggs.

Beginnings of social life

Where, then, are we to draw the line and say that social life has begun? The essence is the idea of reciprocity, that is of mutual help, not only of parent to offspring, but also of offspring to parent. If one female rears and tends her brood, and then they in turn look after their own offspring, this may be a biological advance, but it is not so very different from the Tsetse-fly and other ovo-viviparous insects. One group tend their young externally and 'voluntarily': the other tend them internally and 'involuntarily'.

A more significant advance has been made when the offspring, grown up, help their mother to rear later broods. To be able to do this requires three things: that the original female should live long enough to span the larval life of several generations, perhaps many; that there should be some division of labour, at least between an egg-laying female or '*queen*' and non-egg-laying '*workers*'; and that there should be exchanges of food between the various members (*trophallaxis*) (plate 45).

Associations of mutual benefit sometimes exist between *different species* of animal. The colonies of bacteria or of protozoa that live in the intestines of some insects are examples of this. They help the insect to digest some kinds of food material, especially woody tissues; in return they have a stable and protected environment – at least for a limited period – and a steady supply of fresh material to work upon. Such interspecific associations are called *symbiosis* or *mutualism*.

A social life is an association of mutual benefit between members not only of the same species, but of the same breeding stock. The eggs from which one colony grows up are not necessarily laid by a single queen, but a social colony is self-contained and exclusive. Some colonies deliberately capture members of another species and keep them as slaves, and nests often contain a number of *inquilines*, or 'guests', insects of other groups which live in the nest and are tolerated, or even cultivated as are some of the aphids, for the sake of their products. With these exceptions intruders are at once recognised and attacked, and even regular inmates may be attacked if they depart from their usual behaviour. Even members of the same species from another nest are intruders, presumably because they smell differently. If beekeepers want to amalgamate two colonies of bees they must do so gradually with the help of some device which enables the two colonies to get used to each other's smell, while preventing them from fighting each other.

Fully social insects are found only in two orders, the *termites* of the order Isoptera, and the *wasps*, *bees* and *ants* of the order Hymenoptera.

Isoptera – the Termites

The order Isoptera is a primitive one, well back in the orthopteroid group of the Hemimetabola. Their nearest relations seem to be the rather mysterious Embioptera, which have a sort of rudimentary social life of their own; at least some are strongly gregarious, with wingless females and winged and wingless males, as well as nymphs living together in the protection of silken tunnels. Termites are also closely related to cockroaches, and this is especially evident among the most primitive termites.

All termites are social and live in large, or very large colonies. The individuals of a colony belong to different *castes*, each caste with a characteristic shape, as well as its particular duties in the nest. Some have fully developed wings (*macropterous*), some have short wings (*brachypterous*) or none at all (*apterous*), some are sexually reproductive, some sterile. The remarkable feature of the castes of termites, in contrast to those of Hymenoptera, is that in termites each caste may include both males and females.

Five principal castes are distinguished, three of them sexually active, two of them sterile:

(i) *Macropterous reproductives*, i.e. long-winged males and females capable of emerging on a nuptial flight and becoming the founders, and royal couple of a colony.
(ii) *Brachypterous reproductives*, i.e. sexually active males and females permanently with short wings, which never emerge on a nuptial flight. They supplement or replace the royal pair if necessary.
(iii) *Workers*
(iv) *Mandibulate soldiers*
(v) *Nasute soldiers*
} sterile males or females

In a few species there are no workers, but instead a third sexually active caste without wings (*apterous reproductives*). These may be regarded either as reproductives without wings, or as workers that have developed sexually.

It will be seen from this list that, in spite of being an ancient group with a long evolutionary history behind them, termites have still retained a great deal of biological flexibility. Indeed, the sterility of the worker caste is only relative. Termites are hemimetabolous insects, and so there is no helpless larva which has to be tended. Instead, the young nymph can move about and look after itself. A termite colony, in contrast with a hive of bees, has many more active participants in the work, and successively as the young forms mature they may change their appearance and take on different duties.

History of a termite colony

New colonies of termites are normally founded by pairs of winged reproductive forms – i.e. sexually active males and females – which come out of existing nests in large numbers at certain seasons. This phenomenon is similar to the summer swarming of winged ants in temperate countries (Chapter 17). No one knows exactly what determines the swarming time, nor how it is synchronised among a large number of quite independent colonies. Whatever the stimulus, it involves not only the winged forms but also the workers, which cut special passages to the exterior of the nest, and even help actively as the reproductives issue forth. It is possible that the signal to begin comes from the workers. This is even more remarkable in termites than in ants, because worker ants pass in and out of the nest continually, and so presumably are influenced by meterological and other conditions outside, whereas most termite colonies are cut off from the outside world. It would seem that the influence came from within.

Termites are weak fliers, and their membranous wings are barely adequate for a short nuptial flight. At these times the air is full of them, and they fall an easy prey to many predators. Those few that survive break off their wings, pair off, carry out some form of courtship, mate, find a suitable nesting-site, and begin to construct a hollow nest in wood or soil – but the sequence in which they do these things varies from one species to another. For instance, some mate in the air, but others do not mate until they have jointly built their nuptial chamber.

Termites of the more primitive families live and feed in rotting wood, and when their first clutch of eggs hatches they feed the young nymphs first with saliva, and then with regurgitated food, or with faecal matter. The ability to digest wood depends on the activity of intestinal protozoa, and this way of feeding the young is also a way of inoculating them with these necessary symbionts. First the saliva, which is purely nutritious, then partly digested food, which gradually weans the nymph, while endowing it with the means to digest wood for itself.

As soon as the nymphs can support themselves they begin to look after their parents, and from this stage onwards the royal couple do not feed themselves any longer. Their sole function is of an egg-laying partnership. They remain permanently in the royal chamber, the queen growing bigger and bigger, with a more and more distended abdomen as the eggs mature. This change of shape during adult life is an exception to the general rule among insects, and is referred to as *post metamorphic* growth. It is accounted for by the special circumstances of such an abrupt change from a brief active, aerial life to a long, sedentary one, which may last as long as ten years, and possibly also by the long time that the eggs take to mature (see below).

Like most aspects of the life of termites, the size of the queen and the number of eggs she may lay vary considerably between primitive and advanced termites. Queens of the primitive wood-feeding families become only slightly enlarged, and lay relatively few eggs, perhaps two or three hundreds in a year. This is remarkably few compared with prolific insects of other orders, such as the Blowflies, and it is a tribute to the sheltered life of the termite colony, which must ensure a high rate of survival. Among the more advanced termites both the size of the queen and the number of her eggs are much greater.

The castes of termites

It is important to understand that the termites are a primitive group of winged insects, not far removed from the cockroaches, but much more plastic than these in their behaviour as well as in their shape. In their long evolution they have achieved a great variety of detail within a limited framework. All termites of the present day are social insects, and live in colonies organised on the basis of division of labour and mutual help.

Thus as an order they show nothing like the versatility of, say, the beetles or the flies. When you think of the order Diptera, which includes the mosquitoes, with their very active aquatic larvae, the Sheep Nostril-fly (*Oestrus ovis*) with its larva in the head-cavities of the sheep, and the Cluster-fly (*Pollenia rudis*) with its maggot parasitising an earthworm, the termites seem lacking in evolutionary enterprise.

The flies are remarkable, but not very plastic: that is, with few exceptions, each species follows a fixed way of life, with predictable behaviour. One Frit-fly in a field behaves much like another. Termites and other social insects have much more individuality, and within one species – indeed within one colony – an individual counts for much more. This is contrary to the popular image of an ants' nest or of a termitarium as a crowded community in which all individuality is sacrificed to a common purpose. Perhaps we exaggerate the significance of the apparent numbers when we dig up a termites' nest and see such a milling mass. This is because they are crowded into a small area. Only those who have had to deal with problems concerning infestations of soil or of stored products have any idea of the astronomical number of individual insects that exist in an area of ground. When we are dealing with termites the numbers involved are more obvious.

Be that as it may, the development of an individual termite is not fixed automatically as is that of most insects, but is a result partly of heredity and partly of the treatment it has received from other members of the colony, especially in the matter of food. The combination of heredity and feeding results in certain fairly recognisable forms which we call *castes*, but castes in termites are not nearly as rigid as castes in social Hymenoptera. It is necessary to emphasise this because it is easy to multiply names in an effort to classify all the different forms that have arisen in the various families of termites.

In the first generation of a new colony the nymphs act as workers, feeding the royal pair, and seeing to successive batches of eggs and newly hatched young. The first caste to be differentiated is the *soldier*.

A *soldier* termite, which may be of either sex, is distinguished by its big head, more heavily sclerotised than is usual in the soft-bodied termites, and with small eyes or none at all. Some soldiers have very big mandibles, and are referred to as *mandibulates*, while others have very small mandibles, but the head is drawn out into a snout. These are called *nasute soldiers*, or just *nasutes*. Soldiers are appropriately named because it seems that they do, in fact, defend the colony. Mandibulates use their jaws, and nasutes exude a sticky liquid which traps their enemies, such as ants, when these invade the nest. Apart from these rather spectacular uses of their armament, soldiers often just stand guard in a menacing way, or use their heads to form a barrier, like the shields of ancient warriors. Mandibulate and nasute soldiers never occur together in the same species of termite, but there is considerable range of size in each kind. Sometimes two or even three distinct sizes can be recognised; sometimes there is just a continuous range of size. Soldiers do not feed themselves, even less do they help to feed any other members of the colony. They have to be fed with saliva by the workers, in return for their defensive function.

Most members of a termite colony grow up into *workers*. These look much the same as nymphs, and have wing-pads (unlike soldiers), but never develop wings. Normally they do not become sexually mature, but among these very plastic insects sexually mature workers sometimes occur, especially if all the fully reproductive individuals in a colony have been lost through death or accident.

The workers have plenty of other duties. They do all the nest-building and repairing, carrying the eggs and distributing them throughout the nest, bringing food, masticating it, and feeding it to the king and queen, to nymphs too young to feed themselves, and to soldiers. Worker termites chew endlessly, and everything, wasting nothing.

The digestive economy of worker termites is remarkable among insects of any kind. They start by eating their own and each other's faeces. Wood is a rather unrewarding food material, as so much of its nutriment is locked up, either physically within the cell-walls, or chemically as indigestible celluloses. Even with the help of intestinal protozoa it takes time to extract all that is valuable from the wood, and by passing it through the intestine repeatedly as much as possible is recovered from it.

The more primitive, dry-wood termites eat little other than wood, and it is the workers of these termites which tunnel back and forth within a piece of timber until they have reduced it to a crumbling mass. The more advanced termites live in the soil, and often extend their nests above ground as the conspicuous *termitaria*, or 'ant-hills' of the tropics. These termites have a more varied diet. They eat dead leaves, seeds and so on, which they collect from outside the nest, as well as fungi which they cultivate, perhaps inadvertently, in so-called *fungus-gardens* within the nest. All these foods are collected by the workers, masticated by them, and then fed to all the other castes for whose welfare the workers are responsible. The workers also eat up any kind of debris within the nest, cast skins, dead termites, and the dead bodies of intruders killed by the soldiers. In short, as in Hymenoptera, the sterile workers are the most numerous members of the social community, and are its core in a functional as well as in a numerical sense.

Reproductives, sexually mature individuals, may be of three kinds. The colony is founded by a fully winged pair which may survive for a period of years as its royal pair. Among their offspring appear at times other *macropterous reproductives*, of both sexes, which emerge from the nest with all the other fully winged individuals at a time of nuptial flight. As well as two pairs of functional wings, they have the body relatively well chitinised, and are generally suited to take their place, however briefly, in the outside world of active and competitive insect life. It is thought that they are nearest to the cockroach-like or psocid-like ancestor from which the social termites evolved, and that all the other castes must have been developed from this by specialisation.

If either of the royal pair should die, or if one is experimentally removed from a colony, substitutes are supplied from within the colony. These are individuals which become sexually mature and take over the duties of supplying eggs. They do not leave the nest for a mating flight, and, appropriately, they remain pallid and soft-skinned. The wings are usually only partly grown, making them *brachypterous reproductives*, and rarely they may have no wings at all and be *apterous reproductives*.

How do castes arise?

No one knows the answer to this question. All castes of termites include individuals of both sexes, so that there can be no simple mechanism linked with the sex-chromosomes, or the cell-divisions before fertilisation, such as we shall see later in Hymenoptera.

Three possibilities have been suggested.

(a) *Differences in feeding*

This is a tempting explanation because termites do so much feeding of each other, especially of the newly hatched individuals. This is just the stage at which the presence or absence of particular substances, 'trace elements', vitamins and so on, might have the greatest effect.

It is easy to state this idea in general, but difficult to apply it in detail. For instance it has been suggested that the saliva of the queen may have an inhibitory effect on the development of the sex-glands of those nymphs to whom it is fed. Thus most individuals in a colony would remain sterile. The various partly-fertile ones – the substitute kings and queens, the fully winged reproductives which emerge on a mating flight, and those workers that in emergency can lay eggs – all these could be produced by giving less saliva or none.

Experiments designed to test this theory directly by making extracts from fully winged males and females and feeding them to nymphs were inconclusive. Furthermore, it is difficult to explain how differences in feeding can determine workers or soldiers even in a simple way, and among termites as a whole there are many different shapes and sizes of sterile individuals.

The ultimate difficulty is applying this, or any other theory of caste determination is to know how the colony is able to regulate the relative numbers of each caste, yet it must be able to do this if the colony is to remain a stable community. The limited experiments that are possible confirm that when termites of one or other caste are taken away, so as to disturb the balance, replacements quickly arise. For example, if the royal pair are removed, substitutes are provided by 'promoting' some of the short-winged or wingless reproductives. If soldiers are removed their numbers are restored by the transformation of more nymphs into soldiers instead of into workers.

(b) *The differences are inherited*

This seems less likely than the previous theory. It would require complicated genetics to produce the castes in the right proportions, and there would still remain the problem of producing substitutes when these were needed. Whatever genetical differences there may be between individuals, it is evident that there must also be differences in the way in which they are treated by their parents or foster-parents in order to produce a particular caste when required.

(c) *Mutual stimulation*

The classical example of mutual stimulation is that of locusts which change from the solitary to the gregarious phase when they are crowded together under certain conditions (see Chapter 20). There is some evidence that in a similar way the mere presence of many soldiers in a nest of termites prevents more from being developed until either some of the existing soldiers die, or the numbers of workers becomes disproportionately high.

This is not really a theory but a statement of fact. It is another way of saying that the colony is self-regulating, without explaining how this is achieved. The effect is as if the termites had some level of primitive consciousness, and could take whatever action was necessary to maintain their colony in an efficient state.

TABLE 2
SUMMARY OF THE HABITS OF TERMITES*

KALOTERMITIDAE	TERMOPSIDAE	RHINOTERMITIDAE	TERMITIDAE
'dry-wood termite'	'damp-wood termite'	'moist-wood termite'	'subterranean termite'
← with intestinal protozoa →			no protozoa
no connection with ground; nest within dry wood	usually some connection with ground; nest in damp wood.		nest in ground
Kalotermes *Cryptotermes*	*Zootermopsis* *Stolotermes*	*Psammotermes* *Heterotermes* *Reticulitermes* *Coptotermes*	*Amitermes* *Microcerotermes* *Macrotermes* *Odontotermes* *Microtermes* *Nasutitermes*

MASTOTERMITIDAE consist of a single living species *Mastotermes darwiniensis*, restricted to tropical north Australia. It is a serious pest within the limits of its distribution.
HODOTERMITIDAE are a small family of harvesting termites, feeding only on dry grasses in the open grasslands of Africa and the Middle East.

Termite colonies

Dry-wood termites

The most primitive genus of termites living today is the Australian *Mastotermes*, apparently the sole survivor of a group that was once world-wide. In this genus the tarsi of the legs have five segments, and the wing has an anal lobe, both primitive characters which link the termites with the cockroaches. Also cockroach-like is the habit of laying the eggs in groups, stuck together.

These and other of the more primitive termites (*Kalotermes*, *Cryptotermes*) live in wood, in which they excavate an irregular series of galleries without any definite form, and without any fixed boundary to the nest. No doubt their evolution started in rotting wood, into which it is easy to burrow, and which is already partly broken down by bacteria and fungi. From this some species pro-

* From Harris, W. V., Report of the Seventh Commonwealth Conference, 1960, page 104.

gressed to living wood, and to sound, dead wood forming part of the structure of buildings.

There are therefore the termites which cause most trouble to man, and are the destructive 'white ants' of the tropics. Besides being very numerous, and riddling the wood with their tunnels, they avoid the light and do not normally come out of their nest at all. All their nourishment comes from the woody tissue. This fact makes them even more sinister from a human point of view, because they have none of the exit-holes which reveal the presence of wood-boring beetles, and which would give warning in time for the structure to be saved.

Dry-wood termites that have once infested a wooden structure are likely to live there indefinitely without detection until they have eaten away so much of the wood that the structure collapses. Since they never come out into the open they can reach a particular piece of wood only if it touches something else, either the soil or another piece of infested timber. Attacks usually come to wood that is in contact with soil. Hence in countries where termites are common, all wooden buildings must stand above ground on concrete or metal supports. But the termites have an answer to that. Using chewed wood, saliva, soil, faeces, or any of the various materials they use in their frugal economy, they build a covered passage which runs up the outer face of a tree, a post of a concrete slab, until they reach wood suited to their wants. Thereafter they can move along this tunnel in darkness just as if they were within the nest itself. Those which retain a connection with the ground usually favour damp or moist wood, and belong to the families Termopsidae and Rhinotermitidae (see Table 2). These surface tunnels are a common sight in tropical countries. Owners of wooden buildings need to look carefully for any that run over the concrete base, and which would allow termites to get into the woodwork.

The speed at which these dry-wood termites can excavate their galleries has been illustrated by graphic stories. One, often repeated, tells how during the recapture of New Guinea and other Pacific Islands, crates of stores were left on the ground for only a few weeks, and were so weakened by the tunnelling of termites that they fell to pieces when lifted.

As might be expected when the galleries are untidily arranged and passage to-and-fro must be devious and difficult, the social organisation of dry-wood termites is also primitive. There is no centralised activity, and it seems likely that at times several pairs of sexually active males and females may function within the same colony. They have soldiers, but may have no true workers – i.e. adults that are sexually sterile – relying on the nymphs to feed and tend the colony. This suggests that perhaps the caste system began with soldiers, and that it was only later that some nymphs finished their growth and metamorphosis without turning into either a sexually mature individual or a soldier.

Subterranean termites

If dry-wood termites are sinister, subterranean termites are spectacular. These are more highly evolved genera, with a more closely centralised society enclosed in a clearly defined nest. Characteristically the nest is hidden below ground, in the soil, though some genera, such as *Cubitermes*, build nests in trees, from either a paper made from wood and similar to 'wasp-paper' (see below), or from a cement made from moistening soil with saliva. From these nests covered tunnels run down to the soil.

A nest below ground is usually a complicated but systematic arrangement of cells and galleries, within which can be found a *royal cell* bigger than the rest. The queen and her attendant king remain permanently in this cell, and it is the queens of these termites that become swollen, sausage-like creatures unrecognisable as insects. This enlargement of the body is entirely of the abdomen, the head and thorax remaining of normal shape and size. The sclerites, or segmented plates of the abdomen also keep their normal size, and so nearly all the surface area of the hugely inflated abdomen consists of the intersegmental membranes, stretched, it would appear, almost to breaking-point.

The abdomen of a queen termite may become almost three inches long, and weigh about 1,500 times as much as the rest of the body. Estimates of the number of eggs are necessarily speculative, but queens have been observed to lay eggs steadily at the rate of one egg every two or three seconds. If this rate is maintained she would lay upwards of 30,000 eggs per day, or over 10,000,000 in a year.

One reason for developing this extraordinary, bloated, egg-laying machine is perhaps the slow rate at which the eggs develop. They are said to require from three to twelve weeks. It would be difficult for the workers to store and look after very large numbers of eggs through this period, especially as they would be of all the intermediate ages. If the female holds them in her abdomen for a substantial part of the period she thereby lightens the task of the workers. In general the more primitive the termite, the fewer the eggs, and the less the abdomen of the queen is distended.

The higher termites are spectacular for the way in which they extend their nests up above ground level. The so-called 'ant-hills' of tropical countries are the aerial extensions of the nests of subterranean termites. They vary from a few inches in height to huge mounds of as much as twenty feet, and they are made from earth, wood and any other material that the termites can chew, masticated together with saliva, and which afterwards sets into a cement of exceptional hardness. *Termitaria* of this kind are serious obstacles when ground is being levelled for building or other constructional work. They may resist even modern earth-moving machinery, and may have to be blown up one by one.

Termite colonies of this type have the most elaborate social organisation. The queen lies motionless, steadily laying eggs, and fertilised periodically by the king. The king termite has a continuous rôle, and is not discarded after one mating like the drone bee. Round the queen swarm thousands of workers, who feed, clean and groom her, getting a return bonus of saliva and other exudations which they suck. Workers carry away the eggs as they appear and store them in other, smaller chambers, where they can feed the nymphs when they hatch.

Since these termites are living in a circumscribed nest and not tunnelling into wood that they can eat as they go, they have to fetch food from outside the nest. They get much by tunnelling into the soil, but in some species workers emerge on to the surface and collect dead vegetable debris, or seeds, or even cut green leaves. They come out to forage at night in columns, the workers accompanied and protected by soldiers. In contrast to most termites, which are blind, those which forage in this way have retained well-developed compound eyes.

Vegetable material thus collected is stored away in cells in the nest, and there it becomes mouldy with fungus accidentally introduced with it. Such storage-cells have been called *fungus-gardens*, and it was thought at one time that the termites cultivated the fungus as a source of food. It is believed that they eat the fungus as they will eat anything, but they are not now credited with deliberately cultivating it.

16
Social insects 2: Wasps and Bees

After the termites we have to leap forward to the other end of the evolutionary scale before we find any more insects living a fully social life. These are the wasps, bees and ants, all belonging to the advanced order Hymenoptera.

Comparing these with the termites, two significant differences are at once apparent. The first is that among Hymenoptera the social life is very much the exception. The social wasps and bees, and the ants, attract attention, and stimulate our curiosity, and even entomologists are inclined to forget the many solitary bees and wasps, not to mention the great numbers of parasitic Hymenoptera which have no social life. Out of some six thousand species of Hymenoptera found in the British Isles, fewer than ninety are social, a proportion of about one in seventy. Moreover, among these there is considerable variation in complexity of social organisation. Only the ants, all of which live in social communities, are in this respect as homogeneous a group as the termites.

Indeed it seems likely that the social Hymenoptera are not one evolutionary line but several, and that social life has been independently evolved four or five times. For this to have occurred in one order of insects goes beyond mere coincidence, and suggests that Hymenoptera, or at least the sub-order Aculeata, are in some way predisposed to evolve a social life. This brings us to the second obvious difference from the termites, that the caste system in Hymenoptera is much simpler, with only three castes: fertile females, or queens, fertile males, and workers. All workers are sterile females; no sterile castes are derived from the males. Males, in fact, play no part at all in social life. They exist to fertilise the queen once, and then they die, or are killed by the workers. The picturesque idleness of the *drone*, or male bee is proverbial, and is contrasted with the busy (worker) bee and the industrious (worker) ant.

The determination of sex in individuals, that is which individuals become males and which become females, is unusually simple in aculeate Hymenoptera. Fertilised eggs produce females, with two complete sets of chromosomes (diploid); unfertilised eggs produce males, and remain with only a single sex chromosome (haploid). This is possibly the factor which makes social life particularly easy to evolve in this group, because all fertilised eggs produce females, and therefore almost all the offspring are normally female. If occasionally a few unfertilised eggs are laid to provide males for necessary insemination we have a matriarchal society where large numbers of females can develop the maternal care and sharing of duties upon which social life is based.

This is brought about as follows. A queen receives a supply of sperm from a male and stores it in the *spermatheca*, a single or multiple receptacle found in most female insects. The queen now has control over the sex of her offspring. In the normal way eggs are fertilised on the way down the oviduct, and so give rise to females only. At first she feeds her brood, but soon the offspring take over this duty. Whether the offspring complete their development by becoming sexually fertile, or whether the gonads remain undeveloped and the individual becomes a sterile female worker depends on diet, and for most of the life of the colony nearly all the female offspring do become workers. When sexual reproductives are needed, an augmented diet causes some female larvae to develop fully into young queens. Male reproductives are provided by the simple expedient of laying eggs which have not received any sperm, the duct of the spermatheca being temporarily closed.

A third significant difference from termites is that Hymenoptera are Holometabola, that is their young is a *larva*, quite different from its parent, and requiring a pupal stage in which to complete its metamorphosis. In social insects this larva does not have to find its own food – care of offspring is one of the chief purposes of social life – nor does it help to feed others as do the nymphs of termites. It is, in fact, characteristic of all Hymenoptera except the Saw-flies, Parasitica as well as Aculeata, that the parents lay eggs in a place where the larvae will have an adequate supply of food without having to search for it themselves. Unlike the termites, therefore, the social Hymenoptera have to look after their young throughout metamorphosis.

Wasps, bees and ants differ from each other in so many respects that it is better to consider them separately.

Wasps

The larvae of Hymenoptera other than Saw-flies – of the Hymenoptera-Apocrita, that is – are predominantly carnivorous. The parasitic Hymenoptera carefully lay their eggs on or in the egg or larva of some other insect, or of a few other small animals upon which the larva can feed. The host is not immediately killed, and usually goes about its life for some time, perhaps long enough for the parasite to complete its development, before finally dying of its injuries. This provides the parasite with fresh food continually, but it is at the mercy of the wanderings of its host, which may itself be killed accidentally, or eaten by some other animal, or the parasite may emerge in some quite unsuitable locality. Some of the smallest Hymenoptera parasitise the eggs of other insects (figure 41) and this problem does not arise.

An advance from this is to enclose the host and its parasite in a cell so as to keep them in one place. This usually means making a cell in the ground or in soft wood, provisioning it with food, laying an egg upon it, and closing up the cell. A large group of wasps of the super-family Sphecoidea have evolved to this point, and are known as *hunting wasps*. To feed a hungry larva throughout its life needs relatively large prey, such as one or more juicy caterpillars, and this must be prevented from struggling both when they are

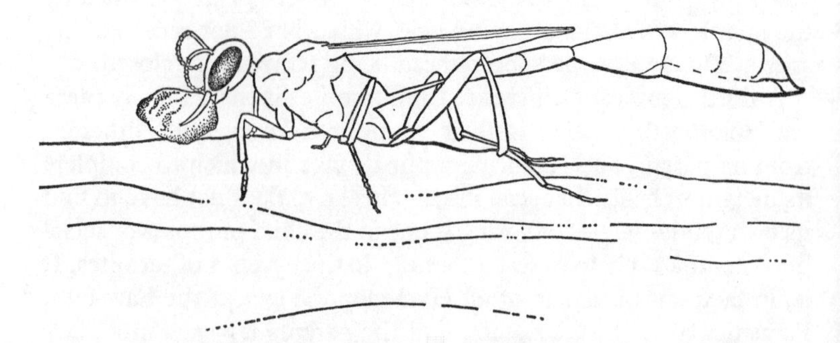

Figure 67. The hunting wasp, *Ammophila*, uses a 'tool' (a stone) to tamp down the earth near its burrow.

being moved, and then they are shut up in the cell. This is possible to the hunting wasps because they have evolved a *sting*, from which the Aculeata get their name, and which is developed from the long ovipositor of the ichneumons. The egg no longer travels down the long valves into the interior of the host, but drops out at the base. The sting consists of the disused valves, which are used to pierce the skin of the prey and to inject liquid from a poison gland. The sting of the hunting wasps sometimes merely paralyses the prey, which then remains alive and fresh, but inactive. Sometimes the prey is killed by the sting, but apparently preserved from rapid decay by the chemicals injected.

These hunting wasps are not social insects because they take no interest in the welfare of their offspring, which they never see after the offspring is grown up. Usually the parent provisions the cell once and then leaves it. Some genera, for example *Bembex*, catch prey that is not big enough for the whole larval period, and they leave the cell open and replenish it at intervals. Such wasps are hard workers, and may go far up into the tree-tops to seek their prey, where no collector of insects working from the ground has ever reached. In several parts of the world collectors have used these wasps to find them rare, and sometimes quite new insects, taking them away from the wasps, which immediately went back for more.

Social wasps

The social wasps belong to the super-family Vespoidea, but even within this group the social species are in a small minority. Indeed many Vespoidea go back to habits more primitive than those of the hunting wasps, and hardly differ from the ichneumons. Various families parasitise different groups of insects, several attacking the larvae of beetles or of Lepidoptera. Mutillidae parasitise the inert pupal stages of solitary bees and wasps, and sometimes other insects such as the Tsetse-fly, *Glossina*. Dryinidae attack chiefly Homoptera. Chrysididae, the *Cuckoo-wasps*, are brilliantly metallic wasps pitted with small punctures, and conspicuous in the sunshine. Their common name is apt because their young are reared inside the nests of social bees and wasps.

The common wasps of the garden and the picnic are familiar examples of fully social wasps. There are several species of the genus *Vespula* in Europe and North America, as well as the notorious hornet, *Vespa crabro*. When these are being discussed it should be remembered that social life among insects probably originated each time in a tropical climate, and therefore social insects that have become adapted to living in temperate latitudes have had to find some way of getting through the winter, with its cold and lack of available food.

The social life of wasps in temperate countries is a summer activity. During the winter the only wasps left alive are fertilised females (young queens), which hide away under leaves or bark, or in some cavity, natural or artificial. They are completely inactive during hibernation, so that not only social life, but any form of life at all is suspended from October until March or April.

The wasps that one sees on warm days in early spring are these queens, awakened from hibernation, and looking for some place in which to make a nest and begin a colony for the summer. This is the time when people believe that if they kill a young queen wasp they will ensure one whole colony of wasps fewer in the summer, but this is doubtful since there is anyway a great natural wastage of young queens. Some are snapped up by birds, others come out too early, on one of those deceptively mild days that are followed by frost, or, perhaps worse, by a period of prolonged cold and damp. Other queens fail to find a nesting-site, since these are not unlimited in any given area. Hence it may be argued that the one or two queens that one can kill at this time of year will not materially alter the number of successful colonies; they will merely reduce the competition. It is a good idea to kill them, nevertheless.

A suitable nesting-place for a queen wasp is in a tree-hole or an underground cavity such as a hole left by a rat or a mouse. Sometimes the queen finds a hole leading into an artificial cavity, such as the roof-space, or under the floor of a hut or outhouse. In any one area there are only a limited number of such cavities that are dry and free from predators which would eat the early nest with its eggs or larvae. So the queen's first big problem is to find a home.

Like the queen termite, the queen wasp at first carries out all the duties herself, and is able to start a new colony without assistance. This is what makes it possible for the wasp society to avoid having to feed workers through the winter, and of course there is then no need to store food in summer for winter consumption. She starts to build a nest from 'wasp-paper' an apt name for what is a sort of plastic wood made by chewing fragments of wood with saliva, spreading the paste and allowing it to harden. The nest hangs from the upper surface of the cavity, and has a central pillar on to which the cells are grafted. Layers of cells are enclosed by *envelopes*, which form a very effective air-conditioned and heat-insulated nest.

The queen makes about ten to twenty cells, open at the bottom and not sealed off like the cells of a honeycomb. In fact the eggs are laid in shallow cells, which are then lengthened as the larvae grow. The queen herself takes only liquid food, chiefly nectar from flowers. In the spring there is no ripe fruit about unless she can find an occasional windfall still lying on the ground, or can raid a human store of fruit. To feed her offspring she pounces upon other insects, bites them to death, and chews up parts of them into a paste which she feeds to her larvae. The full-grown larva spins a cocoon in the cell, from which emerges a *worker-wasp*, a sterile female which is smaller than the queen, especially in the abdomen, but is fully winged, and lives a normal aerial life in the world outside the nest.

Here is a difference from the termites, the workers of which are either immature nymphs, or blind, wingless adults. The worker wasps differ from the queen only in their smaller abdomen and in having no active ovaries. With these they also lack the sexual instincts, and in their behaviour they omit all that is concerned with mating and egg-laying. They take up the behaviour pattern at the point where it concerns cell-building and feeding and care of larvae. They construct additional layers of cells in which the queen lays eggs and soon this is her only activity. Colonies of more than ten thousand workers build up in some years, and these individuals all come from eggs laid by the one queen, and from one batch of sperm received by her at a single insemination.

So through the summer the number of workers increases and they fly out for food for themselves, the queen and the larvae, and tend the larvae, which grow up into more workers. The colony thus grows steadily in numbers. The nest is added to continually, but old cells are also used again.

After the workers have fed the larvae they often lick up a sweet liquid from the mouth of the larva, which is produced by the labial glands. This process of exchanging food is called *trophallaxis*, and is one of the characteristic features of a true social life among insects. The workers do not remove the faeces of the larvae, as do some of the beetles mentioned earlier. The larvae are said to defecate only once, just before they pupate, and the faeces remain in the cell, sealed off by the exuvial of the larvae.

Towards the end of summer the workers build bigger cells, and the larvae in these grow bigger than previously, and develop into young fertile queens and drones. At the same time the old queen begins to lay unfertilised eggs, which are placed in a separate group of cells, and which develop into fertile males. When these males and females leave the nest they pair off and mate, either among themselves or with wasps from other colonies; there is no swarming. The fertilised young queens seek shelter for the winter and begin the cycle all over again in the spring. The males die soon after pairing, and the old colony, with its ageing queen, sinks into decay. Fewer eggs are laid, and many of these are eaten, as well as some of the larvae. When frosts begin, the remaining wasps are quickly killed off and social life comes to an end for that year, with only the hibernating queens surviving.

Bees

One striking difference between bees and wasps is in the matter of diet. Wasps are still chewing insects, and as adults make a great deal of use of their mandibles. Although adult wasps have changed to fruit-eating for themselves, they still give animal food to their young, as did the primitive forms from which they are evolved. Bees, however, have abandoned animal food altogether, and feed both themselves and the larvae upon foods derived from flowers; the protein-containing *pollen* grains and the sweet liquid *nectar*, which the flower produces are a bait to insects. The nectar is partly digested by the adult bee, and is regurgitated from its crop as *honey*.

The bees are thus directly associated in evolution with the rise of the flowering plants, and their structure has been modified to suit this way of life. As shown in (plate 39) the *galeae*, or innermost lobes of the labium have been drawn out to form the long, curved proboscis with which bees can reach into the interior of flowers. Mandibles are still retained, and are used by the leaf-cutting bees of the family Megachilidae to cut out the segments of leaves which they use in the construction of their nest.

Another modification of bees to suit their feeding-habits is the provision of specially fringed hairs to which pollen grains easily adhere. These hairs are specially dense on certain areas called *corbiculae* situated on the hind-legs and the abdomen, and bees that have been successfully foraging can be seen heavily laden with pollen on these parts of the body (plate 27). It is significant that corbiculae are not found on certain bees which live parasitically in the nests of others, and so do not need to collect food for themselves.

Bees are believed to have evolved from ancestors related to the hunting wasps, and to have progressed towards social behaviour in a similar way, first giving their larvae a mass of food at one time. Many species of bees are completely solitary, the social bees being only a small minority. The small burrowing bees of the genera *Andrena* and *Halictus* are common in temperate countries, and are often to be seen going into or coming out of a hole in the ground. There they lay each egg in a cell, which they provision with pollen, mixed with honey from the crop of the adult bee.

Among different species of the genus *Halictus* can be seen differing degrees of maternal care and of progress towards social life. The truly solitary species survive the winter only as fertilised females in hibernation, and in the spring these make a nest, lay eggs, and provision their cells, from which arise one generation of males and females. After a mating flight the males die, and the fertilised females hibernate, to start a new cycle next spring. There is thus only one generation per year. Where the summer is longer and the winter milder two generations may be completed in the year, and some males may hibernate. It seems likely that the ancestor from which they were derived was a tropical insect which produced a succession of generations at regular intervals, and that the habit of hibernation was enforced by the rigours of climate in higher latitudes.

Some *Halictus* make complicated underground nests with branching galleries, and true social life develops when the spring queen lives long enough to continue to lay after the first generation of her offspring has emerged as adults. As she lays only fertilised eggs during this vigorous period the offspring are all female. They have no opportunity to be fertilised since there are no males, so they could produce only male eggs. In fact through the summer they do not normally lay, but spend their whole energies on feeding the larvae that emerge from eggs continually laid by the original queen. Towards the end of summer the old queen and some of the workers begin to lay unfertilised eggs, and so a group of males emerge to mate with some of the females before the winter. The fertilised females can then hibernate to continue the species next year.

These bees, therefore, are maintaining an annual colony much like the colonies of the social wasps, but the individuals are not nearly so numerous, nor are the workers so clearly distinguished from the fertile queens. There are also a number of solitary bees, the females of which have progressed no further than mass-provisioning their cells, and which do not look after their larvae in any way. Nevertheless some of them make elaborate nests, for example the leaf-cutting bees (Megachilidae) which build their

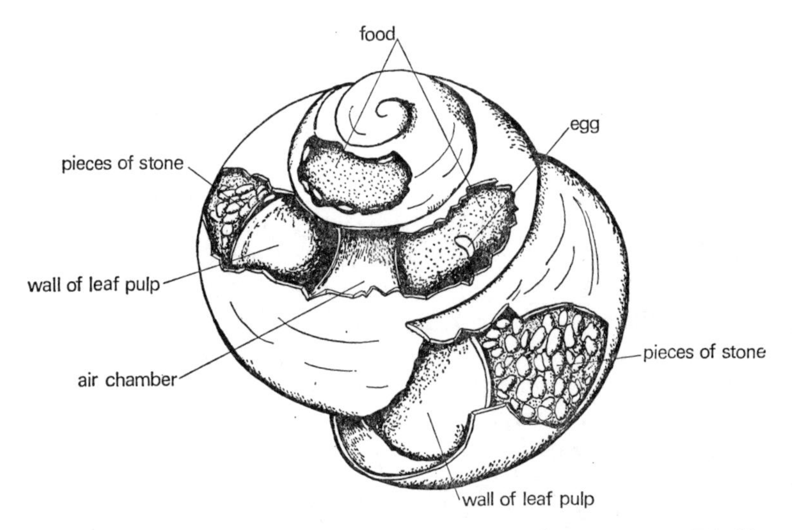

Figure 68. Nest of a mason-bee in the empty shell of a snail (after von Frisch).

cells from pieces cut from leaves. The mason bees (figure 68), *Osmia* and *Chalcidroma*, make a cement from saliva and soil and sand, and construct their cells either in an underground cavity or as an exposed nest against a wall like a swallow's nest. The carpenter bees (Xylocopidae) go further, and use their mandibles to cut a burrow in wood, the chewed fragments of which serve to make a cement from which cells can be constructed.

It is perhaps significant that this active use of mandibles, primitive organs retained from ancestors of chewing habit, is found among primitive bees that have no developed social habits. The social bees have mandibles too, but only as reduced and less effective rudiments. Social bees belong to two groups, the bumble bees of the family Bombidae and the hive bees of the family Apidae.

Bumble bees

These attractive insects of the garden and countryside are also called *humble bees*, especially in books. This has nothing to do with a state of humility, but is supposed to derive from their noisy, humming flight. Most English-speaking people call them *bumble bees*, and this common name is to be preferred because it gives a picture of the endearing way in which these large, furry insects blunder about among delicate blossoms. The English expression 'to bumble' and the German verb 'bummeln' both convey this impression of clumsy, ill-directed activity.

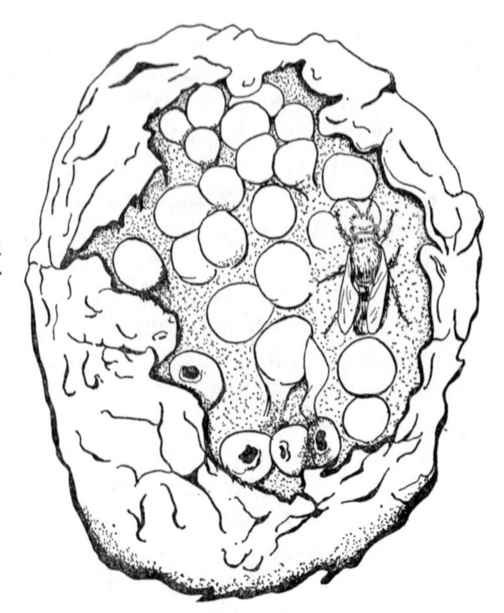

Figure 69. Underground nest of a bumble bee (after von Frisch).

The bumble bees are on a line of evolution distinct from the true honey bees of the family Apidae, and they have some characteristics of the wasps and some of the bees. Like the wasps they cope with the winter of temperate countries by hibernating as fertilised queens, and the social unit with its workers and males is an affair of one summer only. They have abandoned the carnivorous habit entirely, and feed both themselves and their larvae on nectar from flowers, using for the purpose a proboscis developed from the elongate glossae.

The other distinguishing bee-like habit is that of building the cells from *wax* instead of from a vegetable paper. Wax is one of the substances produced in the cuticle of all insects, where it is essential for its property of repelling water. Without the waxy layers, the relatively large surface area of an insect would lose water-vapour so rapidly into the air that insects could not survive as terrestrial animals. A second valuable attribute is that a waxy cuticle is less easily wetted from outside, and the insect has thus a somewhat better chance of escaping being trapped by surface tension on a wet surface, though this is a constant danger to all insects except the very largest.

Insects in evolution have proved ingenious in adapting structures and chemical products of the body for special purposes, and wax is an example of this. Homoptera, notably Aphididae, Aleyrodidae, Psyllidae and Fulgoridae, exude great quantities of frothy wax, which may in some cases have a protective function but which, in the main, seems to be a way of getting rid of waste-products. Bees have developed the production of wax until it provides great quantities of material from which they can build their cells. Of course, they have to make the wax from their normal food materials, and so they have to eat extra quantities of carbohydrates in the form of nectar. Sometimes the bees must lie inert while this carbohydrate is being converted into wax, which oozes between the segments of the abdomen, and is manipulated by the legs and mandibles.

The wax cells of bumble bees are usually built under cover, either under tussocks of grass or, most commonly, in empty holes in the ground that have previously been used by mice. The cell made by the queen contains a number of eggs, and the larvae are provided with honey or pollen as food, which is replenished at intervals. As soon as workers hatch they take over the task of collecting food, and the queen bumble bee never leaves the nest again, but spends the rest of her life in the one task of egg-laying.

The worker bumble bees that are to be seen foraging have a long proboscis and can reach the nectar in certain flowers that are too long for the shorter proboscis of the honey bee. These flowers include the red clover, which is a commercially important crop, and Darwin long ago described the connection between successful pollination of red clover and the numbers of cats and mice, in these words: 'Hence it is quite credible that the presence of a feline animal in large numbers in a district might determine, through the intervention first of mice and then of bees, the frequency of certain flowers in that district.' The workers are very active and collect more than enough food for the young larvae. The surplus at the end of the season is eaten by the young queens in preparation for their hibernation.

Cumber [26] used an ingenious argument to suggest that within the worker caste of bumble bees there is evidence that some members do most of the foraging, while other individuals stay at home and look after the nest. Conopid-flies pursue flying bumble bees and lay eggs on them, which give rise to parasitic larvae. Cumber found that such larvae occurred in about one in four of the bumble bees that he captured as they returned with pollen, but none in the bees that he captured while they were inside the nest working on the combs.

The size of a colony of bumble bees varies rather widely between about one hundred and four hundred workers. In late summer fewer eggs are laid, and some of them are unfertilised. The unfertilised eggs produce males, and these larvae and those from fertilised eggs which are female develop on a fuller diet into sexually active males and females. After a mating flight the fertilised females seek a place in which to hibernate, while the males and the queen and workers of the old colony die from cold and lack of food.

In the tropics there is no need to hibernate, and colonies are not brought to an end by the approach of winter. Young queens may remain in the nest and the colony continues with several queens until it grows too big for its nesting-space. The young queens may leave, but now accompanied by workers, emerging as a swarm to found a new nest somewhere else, like a swarm of honey bees.

Parasitic bees

Bumble bees of the genus *Psithyrus* have a cuckoo-like habit of stealing the maternal attention of a queen *Bombus* (bumble bee). The *Psithyrus* queens enter the nest of a *Bombus* and lay eggs there, leaving these to be reared by the *Bombus* workers. It is not known for certain whether they always kill the *Bombus* queen, but they probably eat her eggs, and they successfully divert the attention of the *Bombus* workers to tending the larvae of the *Psithyrus*. From the nest eventually emerge only males and females of the cuckoo bumble bees.

Psithyrus is adapted to a life of social parasitism, not only in behaviour but also in structure, showing that the habit must have grown over a long period of time. The female *Psithyrus* has no apparatus for collecting pollen, is more strongly built against attack than the workers of *Bombus*, and produces no worker caste of her own.

Stingless bees

There are a group of bees somewhat intermediate in social development between the bumble bees and the honey bees. They are primitive in having retained the mandibles, and indeed have developed these so strongly that they can attack their enemies with them. In consequence they have been able to dispense with the sting, which has diminished into uselessness.

These are tropical bees belonging to the genera *Melipona* and *Trigona*, and are sometimes called the Meliponinae, a sub-family of the honey bees (Apidae); sometimes they are regarded as a separate family, Meliponidae. There are more than two hundred species of stingless bees, and individually they are also very numerous. They produce wax for cell-building, but from the upper surface of the abdomen only, and they mix earth and other substances with the wax instead of using it in a pure state.

The nest is found in a hollow tree, or in some other crevice, or even in a forked branch. Inside the nest are kept stores of honey and pollen, but these are in a space separated from that containing the brood cell. They have queens, workers and males, and like honey bees – but unlike bumble bees – the queen has no pollen-collecting apparatus and does not produce wax. Such a queen is unable to start a new colony by herself, as do the queens of bumble bees and of wasps. This fact has a limiting effect on the habits and distribution, because it prevents them from tolerating a temperate climate by passing the winter as hibernating queens. Bees with this handicap must start new colonies by sending out a fertilised queen accompanied by workers to tend her, that is by sending out a swarm. They must also be able to keep the colony together all through the winter. These are two problems that have been very successfully solved by the true hive bees or honey bees.

Honey bees

Honey bees or hive bees are just a few species of bees that have achieved a most highly organised social life in colonies that are permanent, or at least of an indefinite life. They are of tropical origin, and one species *Apis mellifera* has been domesticated by man, and reared in hives all over the world. The wild honey bees of the Old World tropics, *Apis dorsata* and *Apis florea*, make nests which hang from branches in trees, whereas domestic honey bees and some wild races of *Apis mellifera* prefer to make a nest inside a dark space. In nature, or when a swarm of bees from a hive finds its own nesting place, this is usually in a hollow tree, a crevice or a cave. The bee-hive provided for domestic colonies is a substitute for this, modified in various ways so as to make it easier to handle the colony and to take away the honey that the bees have stored, and which is the main object of bee-keeping.

Like the stingless bees, the honey bees have queens that have no pollen-collecting apparatus on the legs and abdomen, and which cannot produce wax. Hence a colony cannot be started by a single female without help. A queen must go to her new nesting place accompanied by a swarm of workers, who will produce wax for the comb, collect food for her, for themselves and for the offspring from her eggs, and continue to build and repair the honey comb of waxen cells, and performing all the duties of the colony other than the actual laying of eggs.

It is a peculiarity of the honey bee that when the colony builds up to a large population and then divides by swarming it is normally the old queen who leaves, and she takes with her mostly old workers. The young and vigorous members of the colony are left behind. This is a curious arrangement biologically, different from that of any of the social bees. It ensures that the old, well-tried nesting place is not lost, nor do young, vigorous bees dissipate themselves in trying to find a new nesting place and perhaps failing to do so. Yet obviously there will come a time when the old queen will have to be replaced, and this necessitates the process known as *supersedure*, by which the colony rears one new queen in a special cell and after mating she takes over the egg-laying duties from the old queen.

Whether a fertilised egg develops into a queen or a worker is determined by diet. Queens develop from fertilised eggs that are laid, not in the regular worker cells of the honey comb, but in special, large, queen cells that are usually built at the bottom of the comb. Larvae in these cells draw the attention of the workers and receive special feeding from them.

Larvae in all cells of the nest are fed for the first two days upon *bee milk*, a substance produced by the workers from pharygeal glands which open into the mouth. This substance is rich in nitrogen and highly nutritious. From the third day onwards the larvae in the ordinary cells of the honeycomb are given less and less of the bee-milk, and more and more honey, a mainly carbohydrate food. In effect they are undernourished, and they respond by producing at the final moult the smaller body and underdeveloped ovaries that make them into worker bees. The larvae in the queen cells, however, are fed throughout their development on bee-milk, now called *royal jelly*, and in response to the richer feeding they grow a bigger body and fully developed ovaries, and become fertile females, or young queens.

Drones are produced from unfertilised eggs, which the queen lays in special drone cells, bigger than the normal worker cell. The signal to produce drones is normally given by the workers, which build a number of these larger cells in a special drone comb. Into these the queen normally lays unfertilised eggs, producing male larvae, which receive a diet similar to that given to larvae that are destined to become workers, i.e. without the extra protein that goes to young queens. The males are able to become fully sexed on such a diet, presumably because sperm contains only the reproductive cells, without the yolk stores that accompany the eggs.

The production of drones is a piece of communal behaviour that is not clearly understood, involving as it does the workers as well as the queen. It is behaviour that has been perfected in the domesticated honey bees; wild honey bees are less precise in building the various types of cells, and *Apis dorsata* is said to use one kind of cell for either queens or drones. It seems that the workers are stimulated to make drone cells by changes in the behaviour of the queen, and that these changes of behaviour normally presage the laying of unfertilised eggs. This correlated behaviour of queens and workers sometimes gets out of step, and the wrong kind of eggs are laid. The domestic honey bee normally produces many more drones than are needed to inseminate young queens, and these surplus males idle about the hive doing no work, and never foraging, as indeed they have no equipment for so doing. Thus the use of the term 'drone' metaphorically for an unproductive idler is apt.

The honey bee was originally a tropical insect of social habits, with perpetual colonies, and it has met the problem of the temperate winter, not by inactive hibernation of individuals, but by making an artificial environment within the colony. Bees do not fly out in the coldest weather. The little food that they might gather from the few winter flowers would not replace the large amount of energy they would consume, even if they could avoid being immobilised and killed by the cold.

Since in the winter there is so little weather for foraging, and so little to forage for, the bees stay in the hive, and as its temperature falls they cluster together and produce heat by muscular activity, like a mass of commuters stamping their feet on a cold railway platform. In this way they keep up a temperature higher than that outside, perhaps as much as 30–35°C when the outside temperature is down to freezing. This activity needs food, and that is the use that is made of the big stores of honey that bees accumulate during the summer. The whole purpose of bee-keeping from ancient times down to the present day has been to rob the bees of this delicacy, and the useful wax, nowadays replacing it with a minimum quantity of sugar solution to provide them with the carbohydrates needed to keep them from starving.

Man has kept bees from time immemorial, and so we have more knowledge of the habits of *Apis mellifera* than of any other insect. This should not mislead us into assuming that its social life is necessarily of a higher degree of complexity than others.

The evolutionary level of honey bees

The social organisation of a honey bee colony has been carried further than that of a colony of wasps in several directions. One is obviously that the nest is more thoroughly insulated against changes of temperature in the air outside, allowing the colony to live in an artificial environment all the year round. As the bee originated in the tropics this type of nest was probably first evolved to meet the problems not of a winter, but of a dry season when the air was very hot and flowers were few. Inside the tropical rain forest conditions are much the same all the year round, and bees could live a fairly humdrum existence, with no urgent problems, but equally with no great abundance of blossoms, and no strong evolutionary urge. Outside the rain forest, in savannah country, there is greater abundance of flowering plants during the rainy season, but part of each year is hot and dry and food for bees is scarce. The richness of the flowering season could only be exploited by bees if they could find a way of getting through the dry season. It is possible that what to us seems a winter habit of clustering together on a series of closely built honey combs, keeping dry and warm through the winter, was in origin a way of keeping cool and moist during an arid hot season. In both situations of course the bees are not concerned with what is happening outside – their evolution is towards independence of outside conditions, developing habits of behaviour that tend to keep the inside of the hive in the state which best suits the bees.

This problem of how to keep the colony intact through an unfavourable season exists for honey bees and not for bumble bees or wasps because the queen honey bee has lost the pollen-collecting apparatus on the legs, and all the instincts of food-getting and brood-rearing that accompany this. It may be considered an evolutionary advance for the queen to become specialised for a long life of continuous egg-laying, but this carried the grave disadvantage that such queens can no longer start a new colony by themselves. Bumble bees and wasps can survive either a hot season in the tropics or a cold winter nearer the poles, because the fertilised queens can aestivate or hibernate, whichever is appropriate, and start new colonies when good times return. These groups of social insects thus do not have much difficulty in spreading, first to savannah conditions and later into temperate climates.

It looks as if the honey bees developed their form of nest and then spread to savannah country, but that they were carried into colder countries by man. It was early man who accidentally found out that hives of bees could live through the winter, and so he could take them with him into colder countries and use part at least of their winter store of honey as a welcome addition to his own limited diet.

Once a colony can be kept going indefinitely it has a big advantage over colonies that live only for one season. For one thing it can grow much bigger: up to 80,000 has been estimated, compared with perhaps 10,000 in a big colony of wasps. Indeed there is the danger of growing too big, with so many bees foraging that there is not enough food within flying range. The evolutionary reply to this challenge is the habit of swarming. Swarms may have had a dual purpose in the original wild tropical bees, since a shortage of food would occur suddenly as the flowers withered and died, and the dry season approached. A small colony of young bees would be left behind to survive the difficult period of the year, and the swarm led by the old queen would fly to a new area, possibly up in the hills where the hot period was less severe.

Most of the peculiarities of the honey bee can be linked with the food it has adopted, nectar and pollen from flowers in bloom. This food is nutritious, and generally reliable both in quality and in quantity. Of course, the blooming of flowers is greatly affected by weather, so that at times nectar and pollen are plentiful, at other times scarce. Equally there are times when bees can forage for long hours, taking back load after load, while at other times they are kept in the hive by rain and wind. These variations are minimised because the bees store food, and provided that the flowering season as a whole is average, the colony will do well. The store contains food of good quality as well as poor, and the advanced social behaviour of the workers especially their habit of passing food to and fro, gives them all a reasonable share.

A disadvantage of getting food from flowers in bloom is that the bee has to find these. Bumble bees live in relatively small colonies and appear to forage as individuals. They vary in size, many being much bigger than a honey bee, and having a long proboscis that can be used to extract nectar from deep flowers that the honey bee cannot reach. A well-known example is red clover, to the pollination of which the long proboscis of a bumble bee is essential, as Darwin pointed out.

The much bigger colonies of worker honey bees collect food more systematically, and there is no doubt that when a foraging honey bee finds a crop of flowers producing nectar and pollen it goes back to the hive and other workers come out to gather food from the same flowers. The question of how this is done is discussed in Chapter 18 as one of the most complicated examples of behaviour among insects.

Life inside a bee-hive is complicated enough, with the workers building the various types of cells, cleaning, repairing and sealing them, fetching and carrying food in its various forms, wax and propolis, clustering and fanning. Yet each of these activities in itself is a standard piece of behaviour, no more remarkable when considered alone than the behaviour of solitary Hymenoptera when they are preparing and managing a nest. No more remarkable, indeed, than the behaviour of any insect, which must necessarily feed, mate and lay eggs, any of which operations is as complicated as these.

The significance of the behaviour of social insects inside the nest is that instead of each going his own way, the behaviour of different individuals fits together into a pattern, so that together they make up a social community, with a communal life. Yet even so there is something automatic about the whole thing, and an individual in a hive can hardly be said to have much choice about what he does, the colony functioning as a whole.

Outside the hive conditions are different. The bee out foraging is still a worker belonging to the hive, but now she is no longer faced with one of the standard situations inside the hive. She has the much more difficult problem of finding her way to flowers in bloom, loading up with nectar and pollen, and taking it back to the hive. All this needs a different sort of ability from that of toiling in the hive.

For the moment it is enough to note that this is a level of behaviour higher than is shown by any other insect, including other social insects. Even ants, which we shall consider next, and which are highly evolved as individuals, do not have this experience of going out under orders, as it were, to fetch a particular kind of food from a particular place.

17
Social insects 3: Ants

Ants, like termites, live entirely in social communities, and there are many resemblances between the societies of these two orders of insects. The common name of 'white ants' for the termites, though scientifically inaccurate, shows that even to the casual observer the two groups have similarities of behaviour.

A basic difference is that in ants, as in other social Hymenoptera, males have no function other than mating, and the social colonies consist entirely of females; whereas in termites both sexes contribute members to all the castes. Like termites, ants make a mass mating flight, when on a specific day and often at the same time of day, very large numbers of winged males and females emerge from nests over a wide area. The air is full of them, and the newspapers carry stories of 'invasions' by 'swarms of flying ants', reporting this wonder on each occasion as if it had never been heard of before. Colonies are founded by the mated females, which break off their wings. Ants – like termites, but unlike wasps or bees – rear their young not in orderly cells but in an untidy array of chambers and galleries. Their nests are diffuse, with ill-defined limits, and may be in the soil, or above it in mounds of soil, leaves, pine needles and similar debris. Some ants make nests in trees out of 'carton', a substance made by chewing soil with saliva, and more rarely from green leaves. All these nesting habits are strongly reminiscent of those of termites, though true ants make nothing approaching the huge termitaria of the tropics.

These similarities between termites and ants, so far apart in the evolution of insects, are quite clearly to be explained as *convergence*, that is, each group has met similar problems by evolving similar habits. The ant community is organised much more loosely than that of any other social Hymenoptera. Wasps and bees live in colonies with one queen, and the workers follow rather rigid patterns of behaviour in building an orderly nest and in carrying out duties that are fairly clearly defined. The colonies of ants are generally centred round one queen, but sometimes two or more queens are present together, either temporarily or indefinitely. The workers are less closely regimented and can perform a greater variety of acts of behaviour. The expression 'an ant-heap' is often used in human affairs to signify a milling mass of individuals hurrying to and fro each bent on his – or rather on her – business, rather than the orderly coming and going at the entrance to a bee-hive; though ants, too, may follow each other along a trail, and can combine together on occasion with devastating effect.

Ants are believed to be older in evolution than either wasps or bees, and their form of social life is more primitive, and much more diversified both individually and collectively. In particular they are not linked to the flowering plants in the way that bees are. Ants have kept their mandibles, but have added to the primitive habit of feeding on animal food, and most ants will accept almost anything to eat. Ants have abandoned wings, except briefly during the mating flight, and they have become expert runners, covering great distances through and over the soil, and up trees. In the end they can reach any place where any other insect can go on foot, without the hazards of flight, and the great waste of energy that flying entails. In this respect ants resemble the beetles, another ancient and highly successful group, which also has chosen to run rather than to fly, and to use mandibles for a very varied diet.

It is the great versatility of ants that makes them at once so universal and so successful.

Nests of ants

The nesting habits of ants, as of other social insects, are to some extent affected by their choice of food. Ants are thought to have been originally carnivorous, using their powerful mandibles to get the most nutritive food by catching living insects and other creatures. The most primitive living ants are carnivores, and include the fiercely aggressive legionary ants (Ponerinae) and driver ants (Dorylinae). These ants move about in a raiding column, and because of their immense numbers can overcome and eat any animal, however big, that cannot get away from them. In tropical Africa there are many stories of the way in which a column of driver ants passes through a house, killing all the vermin in it. As the column moves it sweeps everything before it.

The workers of driver ants include a wide range of sizes and forms. They are blind, though the workers of Ponerinae have traces of eyes. To get food they march in their terrible columns, either on a raid returning to the same place, or on the move, when they take with them all their brood, carrying eggs, larvae and pupae as they go. Queens are not easy to see, naturally, in such a hostile mob, and not many people care to try. The queens are wingless and march with the rest when moving camp. It is a curious fact that male driver ants are at least as familiar as queens or workers. They are large, winged insects with a fat, cylindrical abdomen, and are the 'sausage flies' which come to light at night and are irritating but harmless as they plop about the lamp.

This peculiarly mobile form of social organisation reminds us of the nomadic hunting stage in man's early history. Such ants obviously do not spend much time or labour in nest construction. They usually crowd into a natural cavity under a stone or under the roots of a tree. When extra cover is needed the ants themselves may link their bodies together. There are periods when the colony raids the neighbourhood and returns to the same nest, and other periods when it moves to a new nesting place frequently. These alternating periods are determined partly by the time needed to feed and hatch out all or most of a batch of larvae, and partly by the amount of food available within a raiding distance of the nest.

This is a primitive type of social life, where the benefits of congregating in large numbers are obvious – a large and regular supply of animal food is given to the brood, and this supply is not subject to uncertainties that affect a carnivorous animal working on its own – but the sheer numbers bring their own problem in rapidly using up the food nearby and so compelling the colony to keep moving to new areas. Even so, ants cannot live in this way unless insect life is very abundant, and so perhaps fortunately driver and legionary ants are confined to hot countries.

The big step forward in the evolution of ant-behaviour came when they broadened their diet and experimented with other kinds of food. It was this more than anything that released ants to become the most ubiquitous of insects. Once they began to eat things that were available everywhere ants could begin to make more permanent nests.

Most of the ants of later evolution, but not all of them, live underground. They do not construct an elaborate artificial nest like that of wasps or bees, but excavate a cavity in the ground by bringing up the soil grain by grain. This is the first of many examples of how ants achieve a considerable result just by the tireless labour of a great number of individuals. As the cavity grows in size the ants leave a supporting pillar, and thus the nest grows into a complicated system of interconnecting chambers.

The eggs and larvae are not kept in individual cells, as are those of wasps and bees, but piled up in some of the chambers. There the worker ants feed and tend the larvae, and frequently move part or all of the brood, either within the nest or, in emergency, outside it. This may be a survival of the primitive habit of carrying the brood along with them, but it has become one of the many acts of maternal attention that are the basis of a complex social life. They have the advantage over bees and wasps that if, for example, the nest is suddenly flooded, or torn open by a predator, the ants can pick up and rescue a great part of their brood.

The nest may be continued above ground by piling up sticks, soil, leaves, pine-needles, etc. into a mound which may reach a height of several feet in the wood-ant (*Formica rufa*) of temperate woodlands. There seems no limit to the number of wood-ants that can find food in a given area: besides the great numbers in each nest, one may find individual nests built against almost every tree in a small area, as in some parts of the New Forest.

Some ants have retained the primitive habit of using natural cavities as nesting places, and many supplement this by making *carton* by chewing soil with saliva. Carton nests may be exposed, hanging from branches, but again the chambers within the nest are relatively large and connected together, and there are no precisely shaped cells such as are made by wasps.

The extreme in nest-making is reached by the ant *Oecophylla*, often quoted as an example of advanced behaviour in an insect. Not only do workers contrive to draw the edges of leaves together, but the ants use silk-spinning larvae as 'needles' to bind the leaves into a nest.

Castes among ants

Basically, ants have the same caste system as other social Hymenoptera: *males*, which appear only briefly for mating; fertile females or *queens*, large, egg-laying machines; and sterile female *workers* which are active members of a colony. The ants that one commonly sees running about are all workers. Ants are nearly all wingless, except for the males and queens up to and including the mating flight.

When discussing any phase of the life of ants it must be remembered that these are a group of great antiquity. During their long evolution, instead of narrowing down into a highly specialised way of life like that of the honey bee, ants have gone the other way in their habits, and become increasingly diverse. Ants go everywhere and eat everything.

This diversity appears not only in their food and behaviour, but to some extent in their bodily shape. Those entomologists who like to categorise everything and invent a name for it have had a happy time with the *polymorphism* of ants, and at last thirty so-called castes have been named. It is a waste of time to bother with these names but those who care to read the account of them in Imms' textbook (page 721) will get a general idea of the sort of variation that may occur.

Which of these shapes is assumed by an ant (or any other insect) depends partly on the genetical structure of the cells and partly on the conditions under which the cells develop into the tissues of the adult insect. The fertile male ant is the best example of a caste that is determined solely by genetical factors. It develops from an unfertilised egg, followed by a larval stage that receives normal food and attention. What develops from a fertilised egg depends greatly on the diet given to the larva, and as in several examples this is how social insects are able by tending their larvae to produce either queens or workers to suit the needs of the colony. The most obvious results of a poorer diet among the workers are lack of wings, failure to develop the ovaries, and a generally smaller body, but sometimes a bigger head.

Among worker ants, however, there are many variations in these respects. Sometimes males and females have short wings or none, and in body structure, especially in number of antennal segments, they tend to resemble the workers. Workers may be very tiny, or alternatively may have big heads and large mandibles, and thus recall the appearance of soldier termites.

Obviously the workers engaged in feeding larvae must be much more subtle in their behaviour than we can appreciate, and can presumably make a considerable difference between one larva and another by varying both the quantity and the nature of the food that they give them. Yet even this is presumably not enough to account for all the observed variations, and it is thought that some hereditary factors enter into the differences between workers. This applies particularly to the big-headed 'soldier' ants.

Then there are pathological forms, in which the effect of either heredity or feeding can be simulated by abnormalities occurring during development. If the cells divide unequally at an early embryonic stage the resulting adult may be a mosaic, with different development in different parts of the body. The best known of these are *gynandromorphs*, which have male characters on one side, or at one end, and female for the rest. The individual may be obviously lop-sided to the naked eye. If some cells show the effects of diet and some do not we may get various combinations of male/female/worker in one individual, and provided the differences are not too extreme, the insect may live to puzzle the systematist.

A number of deformities, some apparently the result of malnutrition, may result from the action of parasites, particularly of the Nematode worm, *Mermis*. This long, smooth worm lies coiled in the abdomen of the adult ant, having grown during the larval and pupal stage of the ant, and taking much of the food that is essential to the normal development of the insect.

Finally a sort of 'pseudo-caste' is the 'replete' of the *honey-pot ants*. An important item of food for most ants is the sugary liquid excreted by aphids (*honey-dew*), and workers often stuff their abdomen so full that the membrane is stretched between the segments. In the honey-pot ants (*Brachygastra, Ptenolepis, Myrmecocystus*) a few of the workers habitually remain in the nest with the abdomen excessively dilated with food, and thus act as living stores, or honey-pots. Other ants treat them as such, and draw from or add to their stores at will.

Of course these phenomena we have just discussed are not peculiar to ants. Gynandromorphs, for example, occur throughout the insects, and are familiar to lepidopterists because the wings of male and female moths and butterflies are often strikingly different; hence a gynandromorph often has a marked difference between the two wings. The effects of malnutrition are less well known outside social insects because among solitary insects a peculiar individual has less chance of surviving amidst the fierce competition of natural selection. It is one of the achievements of social insects that they have turned what is a handicap in other insects into an advantage, and developed it to give them different castes for different jobs in the colony.

The ants have made the greatest use of this individual variability and we are really wasting our time to try to classify all their variations. It is better to recognise the different factors of heredity, nutrition and pathology which influence the shape of the adult worker-ant.

The food of ants

The ancestral ants were carnivorous and, as we have seen, flesh-eating survives as a primitive habit among the ants living today. The legionary and driver ants hunt in packs, and even among ants with a fixed nest there are many that hunt insects and other small animals.

Although very large numbers of insects can exist in a small area, only a proportion of these can be readily caught by predatory ants. Most of these insects are in the soil, in debris on the surface of the ground, or among the roots and stems of grasses. They can be caught by wandering predators, but not quickly enough and in sufficient numbers to supply a large colony of social insects. Presumably this is why the wasps have taken to plant-food for themselves, retaining the flesh-food for the larvae. Some of them have gone over to plant-feeding entirely.

In fact the legionary and driver ants are almost the only social insects to rely entirely on animal food for themselves and their offspring and they can only get enough of it by making ruthless mass raids, which by sheer concentration flush out every insect from its hiding-place. Even so these ants have to move their colony bodily to a fresh area every three or four weeks.

If ants had stuck to a carnivorous diet they would have become stabilised at a small population, much smaller than the numbers of their prey, as is always the case with predators. It was the change to a vegetarian diet, or at least the growth of the habit of feeding on foods other than living animals, that released them from this limitation, and made it possible for them to build up huge colonies, and to spread all over the world.

A first step is petty scavenging, picking up dead insects and other small Arthropods where they happen to be found, and taking them back to the nest. When animal food was plentiful, ants would tackle small prey, or at least insects that an ant could overcome and carry back. It is possibly under the pressure of scarcity, and also when driven to scavenging, that ants first learned to combine, so that several of them could jointly tackle bigger prey. One of the things people always notice about ants is how individuals can drag along objects of all kinds much bigger than themselves, gripping with the mandibles, and thereby converting what was originally a cutting and masticating tool to the quite different purpose of grasping and carrying.

Dragging along an individual load is one thing, but combining with several other ants to push and pull at, say a dead caterpillar, or to dismember one that is still living, is quite another. The latter seems to involve not only a capacity to combine with others for a common purpose, but also some sort of appreciation, however dim, of what that purpose is.

An intermediate level of behaviour occurs commonly when each ant cuts off a part of the corpse (or other food) small enough to become an individual load. This is one of the hazards that face an entomologist collecting insects for preservation, especially in the tropics. Before dead insects can be safely shut up in boxes they must be allowed to dry, at least externally. Otherwise moulds and fungi will soon grow on them. A wise collector will put his open boxes on a table which has its legs standing in pots of water or kerosene, or else next morning he will see a procession of worker ants marching up to his insects, cutting off legs and abdomens – and always uncannily choosing the insects that are rarest and most prized – and carrying them off to their nest.

The scavenging habit is not restricted to animal debris, but is soon extended to plant materials, and especially to seeds. Seeds are a valuable source of food, which not only lies about in abundance on the ground, but by its very nature contains both protein food in the germ of the plant and stores of fatty material provided as nutriment for the embryo plant.

During the long time that ants and plants have been evolving there has been mutual adaptation, and it is suggested that some plants have specially developed the fatty material as a bait for ants, which then carry the seeds about and thereby increase the likelihood of a few of the seeds managing to germinate over a much wider area. Ants store seeds beyond those needed for immediate use, and they have been known to bring the seeds out of store at times and dry them in the sun, throwing away any that have germinated from the damp. Sometimes these seeds give rise to small crop of the plant near the nest of the ants, from which the ants in due course take another crop of seeds. This behaviour has been interpreted as a form of harvesting, and hence such ants have been called *harvesting ants*. It is now thought more likely that the two processes of throwing out the germinating grain ('sowing') and gathering more seeds ('harvesting') are independent, and do not form one sequence of apparently purposive behaviour.

Ants, like termites, have *fungus gardens*, chambers in the nest in which masses of fungus grow, and are used as food. We saw earlier that the fungus gardens of termites may be an accidental contamination, and not be actively cultivated by the insects. No doubt fungi invade ants' nests too, by way of spores accidentally brought in with seeds and other food, or just on the feet of the ants. Yet some ants chew up and prepare a medium from vegetable-tissue and excreta, a miniature compost-heap upon which the fungus can grow.

This is particularly a device used by the *parasol-ants* of South America, notably those of the genus *Atta* and others of the same tribe. Processions of workers carry back to the nest pieces of leaves or flowers, and the little procession has a comic appearance as it wends its way along the ground. Inside the nest this material is masticated and used as a medium for growing the fungi, which are the principal food of the ants. The reasons for thinking that the fungi are actively cultivated by these ants, and not merely tolerated, are that the culture remains pure, and free from contamination with other spores, which must therefore be constantly weeded out by the ants; and secondly that the fungi concerned are botanically peculiar and appear to have evolved in association with the ants. A third piece of evidence is that a queen founding a new colony takes with her a pellet of the fungus in her *infra-buccal pocket*, a useful pouch that ants have beneath their mouth.

A source of food that has been discovered by insects of several orders, but which the ingenious ants have exploited better than any other is the honey-dew of the aphids. As we have seen, aphids spend their lives drawing up the sap of plants through their sucking proboscis. They are small, easily damaged insects, which do not move about much, and for some reason they seem to be compelled to take far more sap than they can digest, perhaps in order to get certain substances that are present in sap in only small proportions. These may be *amino-acids*, relatively simple organic compounds that are the bricks from which complex proteins can be synthesised.

The high proportion of carbohydrates to amino-acids forces upon the aphids this apparently wasteful way of feeding. Ants have discovered this, and busily collect the drops of honey-dew. More than that, they actively stimulate the aphids by caressing them with their antennae. It has been shown that aphids thus 'manipulated' by the ants both suck more sap and pass more honey-dew.

The ants and aphids have clearly lived together in this way for a long time, because they have evolved mutual tolerance. Ants will treat an aphid gently as long as it moves slowly. They will take it into their nest, and there treat it like one of their own brood – provided it does not hurry. If an aphid moves suddenly it is treated either as an intruder (if in the ants' nest) or as prey (if outside) and is attacked.

It is thought that the behaviour of ants towards such other animals as aphids changes during the life of the ant; a process known as *polyethism*. Young workers of *Formica* and *Myrmica* are employed as nurses, and tend both the ant brood and the aphids. As the worker grows older it become first a forager and finally a predator, and then it moves outside the nest, where its behaviour is appropriate.

The relationship between ants and aphids is of benefit to both parties. Honey-dew is in itself a complete food, and some ants can live on it exclusively. It contains sugars, amino-acids, proteins, minerals and vitamins. Moreover the honey-dew that the ants obtain by stroking the aphids is richer in the nutritive materials (i.e. other than the carbohydrates) than the honey-dew picked up by bees and other insects from the leaves, possibly because it passes through the intestine of the aphids more quickly and allows less time for these constituents to be extracted.

The aphid also benefits from the association. Since it has a sort of forced feeding it gets more for itself as well as passing on more to the ant. An even more important benefit is that the ant disposes of the honey-dew, always a problem to the aphid. Such a highly sugary material soon begins to ferment and moulder; we all know what a nasty, sticky mess aphids can make of, say, a tender shoot of *Philadelphus* in the summer. The aphid normally throws the pellets of honey-dew away from itself by kicking with its hind-legs, but the ant obligingly removes the entire pellet. Moreover, the aphids taken into a nest are protected against weather and predators and allowed to live in a controlled environment.

A stimulating account of the relations between ants and aphids is given by Way [116].

18
The behaviour of insects

It is difficult to talk about the behaviour of insects because the very word 'behaviour' has a human prejudice about it. Our own personal behaviour is known to us more comprehensively than that of any other animal, and depends quite as much, perhaps more, upon what is going on in the mind as upon external influences. The *memory* of past events is very important, and much of the human memory has been shown to be subconscious, influencing behaviour according to events that have long faded from the consciousness.

The most characteristic part of human behaviour, however, is a sense of *purpose*, acting according to an imagined situation that has not yet occurred, but which may be as attractive or as alarming as if it had already taken place. These hopes and fears play a great part in human behaviour: have they any counterpart in insects?

Mammals other than man often behave as if they had a purpose. Domestic animals such as horses and dogs, which have lived with man for a long time, have acquired a behaviour that fits in with his. Mostly they do things that they have been trained to do, but sometimes they perform some act that seems to indicate an intelligent understanding of the situation. Although traditionally all other animals are looked upon as being 'lower animals', sharply divided from the human mental level, we instinctively feel that the difference is one of degree only, and allows certain animals to be 'almost human'.

Where insects are concerned the dilemma is more acute. Most of the time they are thought of as lowly creatures, not much better than slugs or worms, but occasionally they are credited with a human, or even a superhuman intelligence. The social insects have always been admired for their busy and efficient way of life, both adjectives which imply that the insects deserve credit for this; that they could, if they chose, be lazy and slipshod. 'Go to the ant, thou sluggard, consider her ways and be wise.'

Before Darwin's day, most writers on natural history accepted the principle of special creation, and wrote about the wonders of insect life as triumphs of divine skill. One easily forgets that Linnaeus and many of the pioneers of the classification of insects, whose work is the basis for our present-day studies, held this view. Johann Wilhelm Meigen, the 'Father of Dipterology', the first entomologist to make a special study of flies, is known to us from an engraving which also bears the biblical quotation: 'God, Thou hast been with me from my youth up; therefore will I show forth Thy wonders.'

The Theory of Evolution by Natural Selection, propounded by Darwin in 1859, was enthusiastically welcomed by many biologists because it suggested a mechanism by which animals and plants had adapted themselves to their environment by a process of elimination, instead of each one having been specially created and fitted into its proper place. Later, the appearance of the work of Gregor Mendel on the mechanism of inheritance strengthened the impression that all biological phenomena would eventually be explained in simple mechanical terms. The behaviour of insects is subject to inheritance and to natural selection just as much as their shape and colour, and so in the present century there has been an overwhelming tendency to seek direct mechanical explanations of the things insects are seen to do. The wonder of the old naturalists at the industry of the bee and the providence of the ant has been replaced by experiments designed to test whether they respond in an automatic way to such influences as temperature and humidity, length of daylight, and direction of the sun.

In science, as in religion, there is always an orthodoxy and a heresy. Down to Darwin it was orthodox to see the ways of insects as manifestations of God's Providence, and heretical to explain them away by mechanics. In the present century it has been more orthodox to explain all behaviour in terms of chemistry and physics, and heretical to attribute any kind of intelligence to insects.

The truth, as always, must lie between the two. An insect is so very much smaller than a man that the surrounding world must look totally different, and a first step in seeking to understand the behaviour of an insect must be to try to appreciate this difference.

Even if the insect were a miniature man it would still find the surface of a gravel path somewhat like a landscape on the moon. The insect has certain advantages that a miniature man would not have: with its six legs and more elaborate feet it can walk more easily on vertical and overhanging surfaces, and of course many insects can fly. On the other hand its eyes are not nearly as efficient as ours, or at least they give much less complete information about the insect's surroundings.

With an effort of the imagination we can put ourselves in the position of an insect in relation to its physical surroundings, but we have no idea what goes on in an insect's mind. The simplest solution is to assume that an insect has no mind, that its behaviour consists only of direct responses to external stimuli, and that the nature of the response is built-in, or 'programmed' by the way in which the insect's nerves are linked to each other and to the organs of the body.

When the relation between stimulus and reaction is simple and direct it is known as a *reflex action;* such is the automatic flight when the tarsi lose contact with the ground. When the automatic activity is complicated, and consists of a sequence of reactions, it becomes *instinctive behaviour*, an example being the elaborate act of egg-laying carried out by an insect that has never seen it done before. Under normal conditions such instinctive behaviour is stereotyped and invariable but both reflex actions and instinctive behaviour can be *inhibited.*

The word 'inhibited' is one that should be used with caution, making sure that it really means something, and is not just an empty name. Thus many bloodsucking insects, such as female Horse-flies, do not attempt to bite until after they have been fertilised. Up to this stage their instinctive behaviour takes them into places where males await them; once they are fertilised their instinctive behaviour takes them into different places, where vertebrate hosts exist, and then further instinctive behaviour causes them to make the motions of piercing and sucking. It is accurate to say that the second sequence of instinctive behaviour was positively inhibited as long as the female fly remained unfertilised, but it is wrong to use the term 'inhibited' in cases where some behaviour merely fails to happen.

A typical example of the empty use of terms is that many larvae which feed in the heads of plants instinctively crawl upwards and towards the light when they are at the feeding stage, and downwards and away from the light when they are fully fed. The first takes them towards the flower-head, and the second towards the soil, where they pupate. To save time and words it may be said that the larvae are at first *positively phototactic* and *negatively geotactic*, and later they become *negatively phototactic* and *positively geotactic*, so long as it is understood that this is giving names and not explanations. They do not move upwards *because* they are negatively geotactic; this is merely another form of words. Accounts of behaviour are particularly liable to offer names for phenomena as if they were explanations.

Learning

Learning is a process by which instinctive behaviour in an individual is modified as a result of past experience. A honey bee recognises nectar and pollen by instinct, and does not have to be taken out and shown by another bee. It may have examined these substances in the hive, and so have been initiated in one way, but the first time it goes to a flower it gathers and carries back the correct material to the nest without receiving any instruction. Once it has found a particular place it goes there again and again until the source is exhausted, whereupon it forages elsewhere. This latter shows a process of learning, and can be imitated in the laboratory, where a bee can be taught to go back to a coloured square or to a pattern of lines on which it was previously given food.

The fact that at least some insects can learn shows that they have a memory. The impression of past events stored in the brain must alter the programme of instinctive behaviour. Conscious behaviour goes a stage beyond this, and introduces a sense of *purpose*. Behaviour is then adjusted not to react directly to the immediately stimuli, nor even to act in the light of past experience, but to achieve some goal, which the insect must therefore be capable of imagining. An example of apparent purpose is a bee leaving the hive on a foraging expedition, with many stops and changes of direction, and which then runs or flies directly back to its starting-point.

To do this looks easy when seen from above, but even apart from the degree of intelligence needed to understand this, the insect is handicapped by its low viewpoint. From down there it is equivalent to wandering round a strange town and then taking a direct line back to the hotel. Many humans could not do this; yet the expression 'to make a bee-line' shows that bees apparently do so.

Primitive and advanced behaviour

This book has stressed the evolutionary approach, and discussed the structure and physiology of insects in relation to a scheme of classification from primitive insects such as Thysanura or Dictyoptera to advanced orders such as Hymenoptera and Diptera. Yet it must be clearly stated that insects of a primitive group are not primitive in all respects, any more than insects of an advanced group are advanced in every way.

Thus the communal behaviour of locusts in a swarm seems highly evolved, and the maternal care of an earwig seems purposeful, and even 'loving' to some of us. The whole order of termites have highly evolved behaviour, far in advance of Crane-flies, though the latter belong to an order that is generally at a higher level of evolution. Examples of different levels of behaviour can be expected to occur in any order of insects.

Reflex activity

The simplest kind of behaviour, as might be expected, is governed by such things as light and darkness, heat and cold, gravity, air-movement and touch or vibration – simple physical stimuli that are always reaching an insect from its immediate surroundings.

Many winged insects, especially those of light, aerial habit such as butterflies, will spread their wings and begin to fly as soon as they lose touch with their tarsi, that is as soon as they no longer feel their weight on their feet. Others, like the migratory locust, are stimulated to fly by the feeling of movement or air past the head. If such a locust is attached by a string to a 'roundabout', that is to a light ring that is free to revolve in a horizontal plane, blowing on to the face of the insect will start it moving, and it will continue to fly until it is exhausted, because it creates its own 'headwind' as it goes.

These simple reflexes, can be inhibited, when the brain calls a halt by overpowering the reflex circuit. Temporary reflexes are common in larval insects, which crawl upwards, and perhaps towards the light, when they are actively feeding, and crawl downwards and into dark crevices when they are fully fed and ready to pupate. These reflexes are of direct value to the insect larva, since the first take it upwards into the vegetation to feed, and the second take it down into the soil to pupate. Natural selection would operate to perfect these reflexes, and to keep them once they have been acquired. Any larva which went the wrong way at either stage would stand less chance of surviving.

The effect of natural selection in perfecting behaviour among immature insects is more powerful than is often realised, as a result of the heavy mortality that is normal, almost all insects that hatch from the egg failing to reach maturity. One female Blow-fly may lay up to 1,000 eggs in its life of three or four weeks. Even though great numbers of Blow-flies sometimes occur locally, the population as a whole does not make any spectacular permanent increase; therefore on average only two (one ♂, one ♀) per thousand of the eggs must eventually produce mature flies. Pure chance plays a big part in deciding which individuals are the 998 that perish, but, even so, very slight differences in behaviour might well increase the chance of any particular individual to be one of the two survivors.

A difficulty about investigating the reflex activities of insects is that they may – indeed, they do – detect and respond to stimuli unknown to man. In light for instance, they react like ourselves by drawing away from the heating effect of long infra-red rays, but many insects are clearly also able to 'see' some of the ultra-violet at the opposite end of the visible spectrum. In the honey bee, for example, it can be shown that ultra-violet light of 3,650 Å has more than four times the stimulating effect of yellow-green light of 5,530 Å, which itself is for us by far the most effective part of the visible spectrum. This, being a difference of degree only from human vision, may not produce any very surprising tricks of behaviour by the bees, though it has often introduced confusion into experiments with coloured discs and flowers.

Bees and many other insects, however, have yet another sense that is quite different in kind from anything human; they can detect the plane of polarised light. This means that they can tell whether the light they are seeing has vibrations in all the possible planes at right angles to its direction, or whether these vibrations are confined to just one plane. This is of very great importance to them, because light scattered from the sky itself is plane-polarised (i.e. its vibrations lie in one plane), and the insects are thus provided with a fixed direction of reference, comparable to the direction of gravity, but quite unknown – at least consciously – in human experience.

Reflex behaviour is not necessarily invariable. Not only may it vary at different stages of development, but it may be different at different times of the day. The honey bee is more strongly disposed to climb up and towards the light in the morning, and down into dark places in the evening. Sometimes the behaviour is altered according to the general brightness, so that many diurnal insects are attracted towards the light in the daytime; whereas many nocturnal insects move towards a light at night. There are seasonal differences in some insects, which do one thing in the spring and another in the autumn, and natural selection ensures that these differences of behaviour are appropriate to the seasons in which they occur. Yet others are related to a general time-sense, or *biological clock*.

It is clear, therefore, that although an insect's behaviour can be analysed, and reflex responses can be detected and studied, the insect is far from being a machine; at least it is far more complex than any machine made by man.

Orientation by light

A *phototaxis* is a movement, the direction of which is governed by light-intensity, and the phototaxis is said to be positive if the insect moves towards the greater intensity, and negative if the insect moves towards the lesser intensity. The light intensity may be perceived by eyes, if these are present, or by the light-sensitivity of the whole surface of the cuticle, as must happen in eyeless larvae. If the light intensity is different on the two sides of the body, the muscular tension, or tone, is no longer equal on the right and left sides of the body. The tone is thus altered in such a way that the insect turns. If the turn is towards the light, and if it continues until the insect is equally illuminated on the two sides, then the insect will finally be facing towards the light, and will continue to move in that direction automatically. If the reflex effect on muscular tone is reversed from left to right, then the insect will turn away from the light, and will then behave with a negative phototaxis.

It may be noted here that turning and movement are not always linked together. The phototaxis is strictly the asymmetrical action upon the muscles, which causes the insect to turn the body in response to the light. If ordinary locomotion follows, the direction of movement will necessarily be governed by the phototaxis, but movement is not an essential part of the reaction.

The above applies when the light source is large and illuminates most of the area of the eye, or of the whole body. The effect of a tiny spot of light at night is different. This is not perceived by blind insects, that is those possessing only a general sensitivity to light in the cuticle. A distant light must be perceived by a compound eye, when it illuminates a few of the facets and not the rest. Most insects that fly at night are said to 'come to light', but they do not do so by turning until the head is evenly illuminated on both sides and then flying directly towards the light.

On the contrary, as soon as the distant light is perceived as a spot of sufficient intensity to produce a reaction, the reflex behaviour is to keep the same facets of the eye illuminated. This is done by flying so that the direction of the light source bears a constant angle to the axis of the insect's body. This may be compared with the light-compass reaction discussed later (see next page). As long as the source of light is distant, the moon by night or the sun by day, the apparent direction of the light is not changed as the insect moves, and so a steady direction is maintained for some time; indeed until some other stimulus changes the behaviour of the insect. On the other hand if the source of light is near at hand its apparent direction will change rapidly as the insect moves. A little thought will show that the result of this is to cause the insect to follow a spiral path, eventually flying round and round the light in ever decreasing circles. 'The moth flying into the candle' in fact approaches in a spiral path, until it is so close that its wings are singed.

If the source of light is not a flame, but an electric lamp, then the insect finally reaches a pool of light in which it is illuminated broadly and the effect of a spot of light is lost. Then normal phototaxis comes into play. Since most nocturnal insects are in fact negatively phototactic they usually fold their wings at this stage and drop down into the shadow beneath the lamp. A few of those which are positively phototactic keep on beating themselves against the light until they are exhausted.

Light traps make use of this reaction to trap night-flying insects. Some, like the Williams Trap, lure the insects in through a series of glass baffles. The Robinson Trap has a series of metal baffles radiating in a horizontal plane through the lamp. A moth approaching on a flight-path that lies exactly in a horizontal plane might reach the lamp, but this seldom happens. If the spiral path is not exactly horizontal, at one point it will dip below the baffles, and the moth will strike one and fall down into the trap.

The existence of this reaction to a light at night is not confined to typically nocturnal insects like moths. Seaweed-flies, for instance, normally fly by day, or at least that is when they are observed, though of course they may be flying out on the beach at night and not be seen. They certainly fly sometimes on warm nights, and then have been found to fly to lighted windows, even to the lamp of a lighthouse as much as sixty feet above sea level.

The Light-Compass reaction

Movement in such a way that a distant source of light is kept on a constant bearing is also found among diurnal insects, and is then known as the *light-compass reaction.*

Beetles, ants and bees have been shown to follow a steady course, keeping a constant angle with the direction of the sunlight. Experiments have usually taken the form either of changing the apparent direction of sunlight by a mirror or other device; or else of keeping the insect in the dark for a while until the sun's direction has changed. The insect usually starts off again at the angle with the sun that it had before being experimented upon.

Yet it is obvious that the light-compass reaction cannot be an automatic reflex, for several reasons. The ultimate purpose of all navigation is to get to a predetermined destination, and the light-compass reaction has survival value to an insect only as long as it leads the insect to some place useful to it. To maintain a constant angle to the sun has no value in itself. Hence something must determine the angle at which the insect starts off, and must modify this angle as necessary to bring the insect to some useful destination. No one seems to have given much attention to the first problem, but a number of experiments show that insects can adjust the angle as the sun moves round the sky.

The light-compass reaction is a pretty device when it is described in textbooks, but the insects clearly make use of it only when it is convenient to do so. They are not controlled by it as slavishly as they are, say, by the tendency to climb upwards, or towards the light, which they will go on doing indefinitely when they are caught in the Malaise Trap. They do not move or fly indefinitely at a constant angle with the sunlight. They appear to use this trick much as humans do when walking over a moor and following an almost subconscious 'sense of direction' making a gradual adjustment to counteract the lapse of time.

Polarised light

Another difficulty about the light-compass reaction is to know what happens when the sun goes in. Some insects certainly cease to move, but others become more active in duller light, and it is certainly not true to say that sunlight is essential for insect navigation.

If there were no atmosphere the sky would be dark except directly towards the sun. All daylight other than direct sunlight reaches the ground as light reflected from particles in the atmosphere. Light waves vibrate in planes at right angles to the direction of movement, and therefore an infinite number of such planes of vibration are possible. Direct sunlight normally contains vibrations fairly evenly distributed among all these possible planes. Light is *polarised* when its vibration is restricted to one plane only.

One of the ways in which light can be polarised is by reflection and the blue light from the sky, which is entirely reflected light, is partly polarised. The proportion of the total light that is polarised is least towards the sun, and in the opposite direction, and reaches a maximum in a ring or 'equator' half-way between these two directions.

The unaided human eye has no means of distinguishing polarised light from that which is unpolarised, and so is quite unconscious of this effect. Insects respond to polarised light, and clearly have some mechanism by which they can distinguish it. It has been suggested that the long cells of the retinula, arranged radially in each ommatidium, may have this property.

No one can tell, of course, what impression the insects get from polarised light, whether it is an impression of glare, or of a distinctive colour. It is perhaps analogous to the ultra-violet also scattered from the sky (and also seen by insects). This is invisible to the human eye, though it produces an impression of glare that may be painful: on a colour film it records as a sort of super-violet, and may extend a picture of a rainbow to eight colours in place of the usual seven.

Bees can detect polarised light, and use it if there is no sunlight. Even a small patch of blue sky is a help to them. The underlying theory is that they are thus detecting, not the direction of the sun, but a direction at right angles to it, i.e. the 'equator' referred to above. They must adjust their behaviour correspondingly.

Individual and communal behaviour

Most insects behave as individuals, and only a limited number take part in group or communal behaviour. This ranges from the dancing of a swarm of midges, or the menacing approach of a swarm of locusts, to the highly organised life of a colony of bees, wasps, ants or termites.

It is a paradox that those insects which behave as a swarm are just the ones which individually behave in the most primitive and automatic way. Perhaps it is not surprising, because human mobs often behave in a more primitive, animal way than the individuals would if they were alone.

A dancing swarm of midges or other small insects is not really a community: it is just a number of insects doing the same thing at the same time. Their behaviour when in a swarm is purely automatic. They have been shown to keep their position in relation to a marker, usually below them, but sometimes above. Thus swarms commonly form above a tree-top, round a higher tower, over a bush, or even over a conspicuous white stone on the bare ground. The strength of the wind must lie between narrow limits, so that they can make slow progress against it until they have passed over the marker, and then allow themselves to drift down-wind again, repeating this process indefinitely. It is this movement of each individual which produces the milling effect one sees in a swarm of midges, which move back and forth as well as rising and falling in the convection currents.

The insects which take part in such a swarm appear to carry out their movements without reference to the others, and a single insect may be seen behaving in exactly the same way. For instance many male insects which mate in the air, e.g. Horse-flies, may be seen in a solitary hovering flight in a clearing or a sunlit spot, and be joined by a female with which after a few exploratory circles, the male goes off to mate either in flight or at rest on the vegetation. It is logical, therefore, to speak of a 'swarm' that contains only one insect. There is some reaction, however, to the sight of the swarm as a whole. A dancing swarm attracts other insects to join it, and in a swarm of locusts the individuals on the edges constantly turn inwards and avoid losing contact with the swarm.

The value of the marker is that it ensures that the swarm – of whatever size – stays in the same locality. If the insects did not keep station by the marker they would inevitably drift downwind, and then might be carried a long way, much farther than normal flight distance, in a series of stages of flight. This happened in 1953–4 when Seaweed-flies (*Coelopa frigida*) were unusually abundant on the south coast of England. Under crowded conditions these flies normally form into a band and migrate along the beach. In that year, under a strong on-shore, southerly wind, many of the flies lost contact with the beach and then, having no ability to find and return to a marker, they were carried away inland and appeared about fifty or sixty miles from the south coast at places as far inland as Oxford.

Though individuals in a swarm may behave independently, they may still be stimulated or excited by the presence of others, and so their behaviour may be intensified, or even changed into something different. Most grasshoppers occur in very large numbers, but the flightless nymphs or 'hoppers' of the gregarious species exert a mutual stimulation on each other which causes them first to march in bands, and then to develop into the gregarious phase of the destructive migrant locust swarms. It has been shown that the change from solitary to gregarious behaviour in these locusts is brought about by the excitement of mutual stimulation, and that this is mainly a tactile stimulus. The hoppers crawl and constantly touch each other, and they can be induced at least partly to change phase towards gregariousness in the laboratory by touching them repeatedly with a needle.

Such masses of insects as the green-fly on a rose, or the caterpillars in an ornamental tree also behave as individuals, and merely aggregate without having any communal behaviour.

Individual insects

The behaviour of the individual insect has been interpreted in two contradictory senses. Experimental biologists assume that insects are mere automatons and devise experiments which tend to show that their behaviour is built up from simple reflexes such as movement towards or away from light or heat, or gravity, and by the light-compass reaction. Yet when we watch, say, a tiger-beetle on its way we instinctively treat it as an individual with some awareness and apparent sense of purpose. It seems to see prey, and to

8*

avoid danger, and appears to show intelligence even if this is of a rudimentary kind.

Of course much has been written about the intelligence of insects that is silly: the fable of the grasshopper and the ant is a good allegory, but bad if it misleads people into thinking that an ant is really busy thinking about next winter and storing up food in anticipation while the feckless grasshopper fiddles away the summer days. But the more work is done on, say, the behaviour of the honey bee, the more it becomes clear that the bee has considerable power of retaining experience in its memory, and remarkable powers of taking short cuts in its behaviour, not only remembering a track and retracing it, following the landmarks of course in the reverse order, but after a time even cutting corners and making what we aptly call a 'bee-line' for home.

The behaviour of bees

When a hive of honey bees is put down in a new place and the exit door opened, worker bees come out, make at first short flights, and then longer ones, into the surrounding country, and soon settle down to their regular job of ferrying nectar and pollen back to the colony. Before the era of experimental science, bee-keepers simply assumed that the honey bee was an intelligent insect. They treated it like any other domestic animal, found out by trial and error what it could and could not be trained to do, and did not bother to enquire exactly how it found its way about.

Experimental scientists start from the assumption that every piece of behaviour has a mechanism for which an explanation can be found in mechanical terms. The honey bee has been subjected to more experiments than all other insects put together – excepting possibly the two prime laboratory insects, *Rhodnius prolixus* and *Drosophila melanogaster* – and while experimental results are frequently clear, and capable of being repeated by other workers, the interpretation of them is much more in doubt.

The type of experiment that has been made on bee's basic ability to find their way about have been on such lines as taking the hive to a fresh area; turning it round or moving it while the bee was away; altering conspicuous landmarks in various ways; taking the bees themselves in small cages to various points and releasing to find their own way back. As in all behaviour experiments, not all individuals behave alike, and so the investigator starts with a prejudice in favour of the 'conformists', and tends to disregard the eccentric bee that does something different. Yet in evolution the eccentrics are the most important, because all evolutionary advances come from them.

The general conclusion is that away from the hive bees depend mainly on sight, and find their way about by remembering landmarks, which can be analysed into colours and shapes. The range of colour vision in bees is rather like the human range, but with less awareness of red and more of ultra-violet. Bees can also detect polarised light (p. 217), and use this as an indication of direction when the sun itself is hidden.

The structure of the insect's eye, as discussed in Chapter 8 suggests that the perception of shape is likely to be more crudely geometrical than is that of the vertebrate eye, which can form a more precise image of its surroundings. The experimental response of bees to isolated patterns shows preference for figures with maximum contrast, and for wheel-like patterns, or those with stripes, preferably arranged radially. Bees also respond best to flickering patterns, and to those with the greatest length of outline between light and dark areas. In short, it seems to be the rapid change of stimulation of adjacent ommatidia which arouses the bees' reaction.

Recognition of flowers

This suggests how bees may find flowers, by being attracted to them not by the beautiful shape and colour that we see, but by the flickering stimulation of the bee's eye as it flies close to a bush in full bloom. It does not explain how they remember individual landmarks by this mechanism, still less how they remember these in reverse and thus find their way back to the hive. Yet an analogy may suggest how this is possible. If one makes a regular suburban journey by train one often arrives without being consciously aware of any of the intermediate stages; yet get on a wrong train by mistake, and one almost immediately feels uneasy, and spots the mistake. It is evident that in fact a sequence of visual and auditory stimuli is normally received and recorded, and any departure from this is at once noticeable. A bee may have some such mechanism which can record and recapitulate the progress to and from the hive.

A bee also uses its antennae, organs which have no exact counterpart in mammals. Near the hive the sense of smell predominates, and bees undoubtedly find their own hive, and detect intruders to it, by the characteristic smell of their own colony. The olfactory cells are particularly numerous on the antennae, which also carry Johnston's organ, a sensory mechanism which responds to bending, and so records such varied effects as the direction of gravity in relation to the bee's body, the balance and orientation of the bee, and the bending of the antennae by wind pressure during flight. The last has been suggested as a way of measuring distance. Bees certainly have a sense of time, perhaps measured by metabolic changes in the food reserves in their body, and so analogous to our own awareness of the approach of lunchtime. Coupling this with the sense of sustained pressure on the antennae which indicates forward movement in flight, the bee may arrive at an assessment of the distance covered.

These senses may give some indication of how an individual bee can find its food and then get back to its hive, and they are probably not greatly different from sense used by all insects to go about their affairs. Bees, however, are remarkable as social insects for their communal behaviour, both inside and outside the hive. Behaviour inside the hive can be studied only by using an observation hive, in which one or more of the combs is visible under glass. The bees seem to tolerate this, but of course, no one can say to what extent they behave differently from normal. It is a remarkable aspect of bee behaviour that in a normal foraging flight a worker bee may fly about in bright sunlight, walk up the alighting-board, and at once enter the completely dark interior of the hive; yet it is able to continue its movements, substituting one set of sensory responses for another without interruption. This alone suggests that insects are not really as dependent upon vision as are mammals.

Most of our knowledge of what goes on in a hive is deduced from its results; for example, the statement that under certain conditions the workers begin to build queen-cells, and to give the enclosed larvae different food which has the effect of rearing them into young queens. One problem that has been studied with the aid of an observation hive is that of how a returning forager tells the others how to find the food – the so-called *Dance of the Bees*.

The dance of the bees

In observation hives, a returning forager bee that has found food

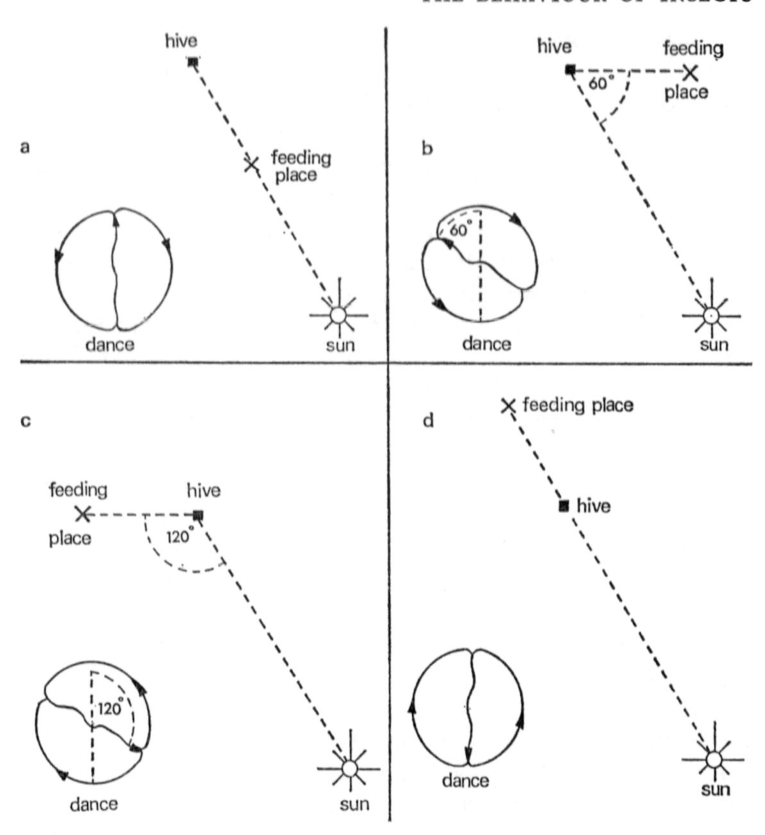

Figure 70. 'The Dance of the Bees'. How the relative position of sun, hive and feeding-place are indicated in the 'waggle dance', or 'figure-of-eight dance' (after von Frisch).

can be seen to perform a 'dance', a sort of gyration on the surface of the comb, which may last from a few seconds to a full minute. While it goes on other workers crowd round and become progressively more excited, eventually following the first worker round and round in a sort of 'follow-my-leader'.

Dances have been seen in observation hives at least as far back as 1788 (Spitzer), but were not investigated in detail until von Frisch. Since then, at intervals, von Frisch has elaborated his observations, and published a summary of them in his book: *The Dancing Bees* [114]. The present account is only a superficial summary.

Although the other workers follow the forager in her dance, they do not issue from the hive in a band and then follow their leader back to the source of the food. If the dance has any function beyond that of generally exciting and stimulating the other workers, it must in some way communicate to them a knowledge of how to find the food supply that the dancing forager has discovered. They then go out to the food themselves.

It is claimed that there are two kinds of dance, a *round-dance* and a *waggle-dance*, with intermediate stages of *sickle-dances* (figure 70). The first indicates the existence of a food supply, but gives no indication of direction, and relates to a source relatively near the hive. The waggle-dance can be orientated on the comb so that the imaginary line drawn through the centre of the figure of eight has any direction. Observation suggests that round-dances are performed by workers reporting food at distances up to about fifty yards, and waggle-dances by those reporting food beyond one hundred yards, with a mixture of the two kinds of dance-pattern for intermediate distances. Experiments with a number of dishes of sugar-solution, all at the same distance, but in varying directions, indicate that bees attracted out by round-dances stay within the radius of fifty yards, or a little more, and forage in all directions; whereas those stimulated by waggle-dances not only fly further before searching, but go off in approximately the right direction, and do not waste effort in other directions.

It is also claimed that the angle that the correct direction makes with the sun is, within a reasonable error, the same as the angle that the axis of the waggle-dance makes with the vertical on the comb (figure 70). This is perhaps the most difficult part of the mechanism to believe. The bee flies out in a general hunt for food, finds it, returns directly to its nest, takes note of the apparent direction of the sun looking back, as it were, over its shoulder, and then converts this direction into an angle with the vertical and proceeds to indicate this on a vertical comb, in complete darkness, and not even by taking up a stationary position, but in a figure-of-eight dance, in constant movement.

The conversion of horizontal direction in relation to the sun into a vertical direction in relation to gravity is perhaps not as incredible as it might seem to the human mind. The insect's senses are more independent of each other than are those of vertebrates, and a simple substitution of one sense for another is more easily achieved. There is some evidence that this may occur in ants, which do not seem to make any use of it. The 'cleverness' of the bees in this respect seems therefore to lie not in their ability to do this, but in their practical use of the trick.

Assessment of distance

Beyond the minimum for the waggle-dance to occur at all, distance is said to be conveyed by timing. As distance *increases*, the number of circuits of the dance per minute *decreases*, but the length of each run during which the abdomen is waggled *increases*, and so do the number of waggles. So besides helping to distribute the scent of the food to the surrounding workers, the waggles are considered to be a time code.

The qualifying words 'is said to', 'seems to' and so on are necessary in reporting this mechanism because the dances take place quickly, and the pattern is not constant. That they occur, and that they are some form of communication, is not to be disputed, but the patterns shown in the diagrams may be no more than inventions of the observer. When it has been agreed that the bees might make such a dance as a code, it remains to be explained how the surrounding workers can appreciate it, in the dark, and standing on the same level as the dancer. They do not have the advantage of seeing the dances in plan, from above, and in a good light, dotted lines and all. They get only momentary contacts, comings and goings in a very confused sequence.

Wenner [118a] has put into words the doubt felt by many people that the surrounding worker bees can possibly read as much into the dance by visual observation as is claimed for them. He points out that those workers that receive a message from the dancing bee, and fly off to the food, are those which actually touch the dancer with their antennae. Moreover they do not touch the waggling abdomen, but the thorax.

Wenner discovered that during the straight part of the dance, when the bee is waggling, she is also emitting a sound of a frequency of 250 cycles per second. He suggests that the duration of this vibration is an indication of the distance of the food, and is perceived by the surrounding workers through the antennae. It is agreed that bees have a capacity to appreciate the length of a period of time, since they can be trained to come to food at regular intervals. It is therefore not too great an effort to the imagination to believe that the returning forager can be aware of the length of time it takes to fly back to the nest, can indicate this by the length of buzzing time on the straight part of the dance, and that other bees can reverse the process and fly out for a similar length of time.

The bee is thus giving an indication of time of flight rather than of distance, and this is only reliable if bees in flight maintain a fairly constant speed over the ground, whatever the speed and direction of the wind – provided of course that its strength is not beyond their powers of flight.

In this account Wenner still allows that direction in relation to the sun may be indicated by the orientation of the dance on the comb in relation to the vertical, but he adds the significant comment: 'It may conceivably turn out that the foraging bee's entire message is conveyed by sound signals.' It may, indeed.

Even if a sufficient number of workers follow the forager in the dance, and so get its pattern imprinted on their senses, it is still surprising that they can do so with sufficient accuracy, as the dance is on such a small scale. Compare the size of the dance of one bee with the ship's compass of an ocean liner! Though they pick up with their antennae the smell of the food, and only need to navigate to within smelling distance of the plant to find it, their performance is still a remarkable one.

Changes of direction

A further complication arises when the food supply cannot be reached by flying in a steady direction, but the bee must fly round an obstacle such as a hill to get there, and back. Von Frisch claimed (1966) that the dance then indicated the distance and direction *in a straight line*, a line upon which the pioneer forager had never flown, and which could not be taken by her followers. This result, if experimentally correct, indicates a power of abstract thought that would be beyond a great many human beings. This does not make it untrue, but it does mean that the mental equipment of a bee must be assumed to reach a formidable level of complexity.

The mental level of bees

To summarise, therefore, the dances of the bees seem to lead unavoidably to the conclusion that bees have mental equipment of a high order. The simplest course, and the traditional one, is to say that they are 'clever' or 'intelligent', and can do whatever is necessary to deal with the problems of their everyday life. It is considered heretical to use such adjectives about insects; we must believe that the responses are mechanical, and invoked by physical and chemical stimuli that can be isolated and studied. If we do this we are led back into the complexity of inventing reflexes which are presumed to reinforce and replace one another as the situation changes. Finally we are back where we started, saying simply that bees can do certain things, and do them as they may be necessary. And that is as far as we are likely to get.

19
Insects against Man and his animals

The relations between insects and Man can be looked at from two opposite points of view.

From the human viewpoint, insects are noticed if they are a nuisance because of their buzzing, settling on the body, or on food, and in particular if they bite or sting. From ancient times they have been suspected of carrying disease, and Man has made a half-hearted attempt to drive them away, by using fly-whisks and insect-powders, and to avoid them by taking himself and his domestic animals and herds to other areas.

No systematic or coordinated war on insects has been carried out until very recent times; only in the last few decades, in fact, when large-scale agriculture, and rapid communications between different areas of the world, have men been forced to pay attention to the activities of insects.

Those insects with which Man is thus concerned are few, and those insects which Man has found useful are even fewer. The honey bee, the silkworm, the lac insects are cultivated for their by-products, and a few insects have been eaten as food. The principal direct benefit derived from insects in general is aesthetic pleasure. Butterflies are admired by everyone, and the summer scene would not be complete without the presence and activity of many other insects.

The relationship takes on a different aspect when seen from the viewpoint of the insects, which have been in existence a great deal longer than Man. Even in late Devonian times, about three hundred million years ago, insects had clearly existed for some considerable time to have reached the state of evolution of the fossils of the succeeding Carboniferous period; whereas Man has existed at most since Pliocene times, one million years ago. There are at present at least one million species of insects, and countless millions must have existed at one time or another. There is only one species of Man, and even if his nearest relatives are included there can never have been more than a handful of them.

In numbers of individuals, too, the insects dominate the zoological scene. A single bee-hive, under normal conditions, may contain as many as 50,000 to 80,000 individuals. A single swarm of locusts may cover eight square miles, and involve an estimated thousand million individuals. The numbers of tiny flies in swarms almost pass belief. It has been estimated that there might be 3×10^9, or three thousand million, larvae of midges in one lake, and that several hundred million adult midges may emerge from one stretch of the River Nile in a single night.

Compared with these figures, human beings are few, and all our talk of 'explosive' outbursts of human population in the present century are foolishness. Moreover, any such outbursts are very recent events in evolutionary history. Man's problem has been under-population, how to feed and breed enough. The present alarm at over-population is that Man wants to maintain a certain standard of living, and is not content to survive as a species by the zoological method of waste, each pair breeding enormous numbers of offspring, of which all but two, on the average, will perish before they grow up.

The expression 'insect enemies of Man' is a misconception. The insects that have a direct relationship with Man, as one species to another are very few indeed. The Human Louse, *Pediculus humanus*, is just about the only one. The Congo Floor Maggot, the larvae of the fly *Auchmeromyia luteola*, may be another, but the fact that this fly is localised to one part of one continent, Africa, suggests that its habits may have been developed comparatively recently in human evolution.

All other relationships between insects and Man are accidental. The fact that an insect injures a man or a domestic animal, or eats a cultivated plant, instead of a wild one, is not of biological significance. Hence the insects with which Man is concerned are not by any means a natural group, but an assortment of casual species taken from any or all of the various orders of insects.

Nature of injury to Man

Insects injure Man and his interests in the following ways:

(a) *Stinging* Although the popular idea of any insect is that it might sting, this is actually a very limited form of attack, confined to the aculeate, or stinging, Hymenoptera, bees, wasps and ants. In these insects the valves of the ovipositor have been relieved of their original purpose of guiding the egg into a remote place, and have been developed as a weapon of attack and defence. Wasps, which feed their young on animal food, use the sting to paralyse or kill other insects, and do not normally use it as a fighting weapon. They can do so if provoked, however, and it is the intense pain and dangerous inflammation brought about by the stings of wasps that are most feared.

Bees have given up animal food for their larvae as well as for themselves, and their sting is purely a weapon. Their social life is intensely exclusive, and they sting any other insect (with a few privileged exceptions) that comes into their hive, even bees from another colony. The barbed stings of bees cannot be withdrawn quickly from the human skin, which is more rubbery and retentive than the integument of another insect. The sting is torn out, together with its sac and adjacent tissues, and the bee dies of the injury. Clearly the sting of bees was not evolved as a weapon against large vertebrate animals.

Ants are the most versatile of all insects, and their sting is employed whenever it is useful, for killing prey or for repelling enemies. In either event the cooperation of a large number of ants enables them to sting to death prey much larger than themselves. Thus the dreaded Driver Ants of the tropics will kill any animal, however large, that cannot shake them off and get away.

(b) *Biting with mandibles* Insects rarely bite Man in the strict sense, that is by using their mandibles to do so. Mandibles are used for chewing vegetable and animal food, and for attacking and devouring other insects, e.g. the carnivorous ground beetles, tiger beetles and the voracious Dytiscidae and other water-beetles. Some of the biggest stag-beetles can grip the human skin with their mandibles – or at least the females can; the mandibles of the males are usually too long and clumsy to do this – but the bite is little more than a pinching of the skin. So although many other insects bite other insects, even catching them in the air as do the Dragon-flies, none, not even the so-called 'biting lice' of the order Mallophaga, regularly bite vertebrate animals by using the mandibles in opposition to each other like jaws.

(c) *'Biting' by piercing and sucking* Insects that 'bite' Man nearly always pierce the skin and suck blood, either directly from a capillary, or from the pool of blood that escapes from the broken vessels. The evolution from mandibles acting in opposition into mandibles which combine with maxillae to form a *fascicle*, or bundle of piercing stylets, is a step that has occurred at least twice in the evolution of insects. The first time was among Hemimetabola, giving rise to the Hemiptera (Homoptera+Heteroptera), the sucking lice (Siphunculata of Anoplura), and one or two small orders; the second time was among Holometabola, giving the flies (Diptera) and the fleas (Siphonaptera).

Though this type of piercing proboscis (see figure 53) is sometimes employed to suck the body fluids of other insects it was undoubtedly perfected in evolution chiefly as a way of piercing any relatively large surface to get at an abundant source of liquid food. Hemiptera evolved it first to suck the sap of plants, a more economical way of feeding than to chew an immense amount of cellular plant-tissue, as do the grasshoppers and locusts, and the caterpillars of Lepidoptera. This type of feeding reaches its highest development in the aphids and other small Homoptera, which feed and breed continuously in the summer. A few bugs such as *Rhodnius* and *Triatoma* have used this equipment to pierce the skin of warm-blooded animals, and the Siphunculata, or sucking lice have become wholly committed to this diet, and parasitic in consequence. The Human Louse is just one species that happens to have evolved with Man as its host.

Primitive flies devised this method of feeding as the normal among females, which needed more protein to mature the eggs. It is thought that all primitive bloodsucking flies – i.e. the mosquitoes, Black-flies, biting midges and Horse-flies – evolved from one bloodsucking ancestor. Tsetse-flies and Stable-flies, in contrast, arose from a group that had already lost the mandibles, and so had to adapt the labium before they could pierce and suck.

All bloodsucking flies of the primitively mandibulate group remain free-living: i.e. they do not attach themselves to one host, but fly about and take 'blood meals' as they need them. They thus attack any vertebrate animal that occurs in a suitable setting, or behaves in such a way that they identify it with food. They do not specially choose Man or his domestic animals.

Fleas on the other hand are blood-feeders in both sexes, and spend their whole adult life sucking blood from a host. They are therefore described as ectoparasites, though they are not confined to their host as is a louse, and they can live away from their warm-blooded host for long periods. Although fleas are associated with particular hosts, and are known as the Cat Flea, the Dog Flea, the Hedgehog Flea, and so on, the adult fleas will bite any host that they can find. The specificity implied by the name is governed by the habits of the larva, which feeds on debris in the nest or den of the host, and it is to this that the fleas are specifically adapted.

(d) *Invasion of tissues by larvae* The nymphs of hemimetabolous insects are generally so much like their adults that they have similar feeding habits. Larvae of Holometabola, in contrast, are quite different creatures from the adult insects, and are not in any way limited to the same kinds of foodstuffs. Whether or not they attack the body of Man or of his domestic animals depends on the mouthparts with which the larvae are equipped, and upon their taste in food.

As a rule, mandibulate or chewing larvae have little or no opportunity to bite Man, or to feed on his body tissues. Occasionally they may be swallowed. This happens sometimes to larvae of *beetles*, and to the caterpillars of small *moths*, either of which may be infesting human food. Infestation of the human body by beetles is called *canthariasis*, and by caterpillars is called *scholechiasis*. These terms are rarely used in practice, and they are given here mainly to emphasise that *myiasis*, a term that is often used, can strictly be applied only to attack upon or invasion of vertebrate tissues by the larvae of true *flies* of the order Diptera. Myiasis is by far the most important aspect of personal attacks by larvae.

Larvae of any flies may be swallowed accidentally, and may or may not survive in the intestine and cause damage there. The *maggot*, the general-purpose larva adopted by all higher flies of the sub-order Cyclorrhapha, is equipped with mouth-hooks, which are usually curved, and often pointed and very strong. They can be used either for guiding liquid of soft pulpy materials into the pharynx of the larva, or to pierce tissues of plants or of animals.

Maggots are usually either *amphipneustic* with a pair of spiracles on the prothorax and another at the tip of the abdomen, or *metapneustic*, with the posterior pair only. In either event, the larva is able to immerse itself almost completely in the body fluids or tissues of its victim, and to feed continuously with its mouth-hooks, while exposing only the small area of the posterior spiracles to the air for breathing purposes.

Thus the maggot, which in most flies feeds in decaying vegetable or animal matter, including dung, can take to feeding on the mucus of the nose or throat, on pus from wounds and sores, on dead or diseased tissues, and finally on healthy living flesh.

Sometimes this ability is, as it were, an acquired taste, and additional to the normal range of feeding. Greenbottles of the genus *Lucilia* are abundant in all countries, and sometimes reach fantastic numbers. Their normal breeding material is carrion, organic matter of animal origin, and they are found in great numbers in dustbins, refuse-dumps, and round abattoirs and anywhere where scraps of meat or fish are to be found. The larvae of certain species including the cosmopolitan *Lucilia sericata*, and the more restricted *L. cuprina*, live in the soiled wool round the anus and crutch of a sheep. They feed on the products of bacterial decay, but they also attack and enlarge any wound or sore in that place, and can lacerate the healthy flesh to start a suppurating wound on which they then feed. Hence these species of *Lucilia* are spoken of as 'sheep maggot flies'. Some other *Lucilia* attack the mucus in the heads of toads, and eat great cavities before finally killing the toad.

A number of other Blow-flies and their relatives invade human tissues, and those of domestic animals. Some *Cuterebra* are *facultative carnivores* resembling *Lucilia*; i.e. they can feed on living flesh if they get the opportunity, and on dead flesh (carrion) or on dung as a regular thing. Some other species have become *obligatory carnivores*, and never feed except at the expense of a living animal. These species can be hideously dangerous to Man, because they lay eggs in any exposed wound or sore, or even in a running nose, especially when the victim is asleep. The maggots may feed for a time without being detected, concealed in the copious flow of mucus, which they stimulate by scratching with their mandibles. As they grow bigger, however, they begin to eat the living tissues of the head or face, until the victim is permanently disfigured – or even killed.

Compared with these, certain obligatory parasites are minor and local in action. The Tumbu-fly, *Cordylobia anthropophaga*, is an African Blow-fly the larva of which lives in a boil or swelling, feeding on the tissues, blood, serum, lymph, while breathing air through its hind spiracles exposed through the centre of the boil. The larva of *Stasisia rodhaini* has similar habits in ungulate mammals, though rarely in Man. In South and Central America the Human Bot Fly, *Dermatobia hominis*, lives as a larvae beneath the human skin, as its name implied. As was explained with admirable clarity by Bates in his account of his life on the Amazon more than a hundred years ago, the flask-shaped body of this larva makes it extremely difficult and painful to remove.

In spite of its name the Human Bot Fly is also a great pest of cattle, and indeed none of these larvae are specific to Man. *Cuterebra* and its relatives are believed to have originated as parasites of rodents and to have spread to Man when he associated with rodents, however inadvertently. Only the Congo Floor Maggot, *Auchmeromyia luteola* appears to have evolved its bloodsucking habits in direct association with Man. This larva sucks blood from the bodies of men and women who sleep on a mud floor, and is unable to climb up into even a low bed. Even here it is not Man's specific person that is the focus of evolution of the parasite, but his habitat.

Domestic animals are much plagued by invading larvae, the best known and economically most important ones being the Warble-fly of cattle (*Hypoderma*), and the Stomach Bot of horses (*Gasterophilus*), and Sheep Nostril-fly (*Oestrus ovis*). Camels have a throat bot, *Cephalopina titillator*, which when fully fed is expelled by sneezing. This is a spectacular parasite, which may contribute to the haughty carriage of the camel's head, but which is of little practical importance.

Even the big mammals such as rhinoceroses and elephants suffer from myiasis. A rhinoceros's stomach may contain two or three hundred larvae of *Gyrostigma*, all attached to the stomach wall, and feeding on the food that passes through. Since they do not in fact attack the tissues of the host, but merely rob him of some of his food they are sometimes called 'kleptoparasites'.

Elephants, too, have stomach bots (*Cobboldia*), *Ruttenia* in the soles of the feet, and a skin warble, *Elephantoloemus indicus*. The last is of practical importance in the domesticated elephants of the

Indo-Malaysian countries, and can cause serious loss of condition if it is not given early treatment.

(e) *Personal nuisance by insects* The insects that cause a nuisance are so many that it is impossible to catalogue them. Except for the butterflies and the few beautiful beetles and Dragon-flies, the nuisances are the only insects to attract peoples attention. Those which are a menace as well as a nuisance are mentioned in other sections and chapters, notably bloodsucking insects above and locusts below.

Individual insects are a nuisance when they crawl over one, but the most persistent are the swarms of flies which gather round one's head and face. Some of these are attracted by the perspiration, and waving one's arms and beating at the flies only makes one even more attractive to them. The muscid flies of the genus *Hydrotaea*, of which *Hydrotaea irritans* is so aptly named, are those which pester one on country walks. They, again, are not specific to Man, but make the lives of cattle a burden by blustering round the muzzle and eyes for the sake of the liquid secretions there. The sweat bees (some species of *Halictus*), and some of the Meliponine 'stingless bees' cluster on the hands, and can irritate the skin by scraping it with their mandibles.

The clouds of midges (Chironomidae) that often attach themselves to a person, and move along with him, are merely interested in him as a prominent object in the landscape, and are using him as a marker for swarm orientation.

Transmission of disease

Any disease that is not a result of malfunctioning of the body itself is caused by the activity of a 'germ', some parasitic animal or plant that lives at the expense of the host and causes a disturbance of his physiology. Such organisms can be passed from one host to another through the agency of an insect, in three ways.

(a) *By contagion* The intervention of insects in contagious diseases is no different from any other source of contagion, infected food, dirty cutlery and utensils, air pollution, or contact with the body or clothing of an infected person. Insects crawl over an infected person or a soiled object, and fly or crawl to the person, food or property of someone else hitherto untouched.

Almost any insects may be involved in an infection of this nature, but some are particularly suspect because of their habits. The most notorious suspect is the House-fly, *Musca domestica*, which not only breeds from preference in human refuse, and the dung of man and his domestic animals, but also as an adult lives in his house and crawls over his skin and his food.

The padded feet of the House-fly, which are shown in great magnification on the anti-fly posters, are not different from those of most flies but the habits of *Musca domestica* make it certain that it must inevitably carry many bacteria, viruses and other organisms of disease about. Yet controlled experiments have never succeeded in showing that House-flies are the sole cause, or even the vital factor in any human disease. They certainly help to spread various enteric infections, and have a link with poliomyelitis, which will need careful investigation. One can say that if there is contagious disease about, then House-flies will help to keep it circulating, and on that ground alone they should be swatted.

The other flies that are particularly liable to spread disease by contagion are the 'Eye-flies' or 'Eye-gnats', *Siphunculina* in the Old World and *Hippelates* in the New. These are flies of the family Chloropidae, the adults of which are attracted to suck the mucus round the eyes, nostrils and mouth. Individually they are tiny and insignificant. Their menace lies in the impossibly large numbers which make it statistically certain that any contagious organisms will be quickly distributed. Conjunctivitis and streptococcal infections of the eyes and lips are the most prevalent infections to be spread in this way.

On a smaller scale, the domestic cockroaches crawl about over human food, or fall into the soup. Apart from the unpleasant smell, and the nasty furtive way in which they scurry away when the light is switched on, there is a risk that cockroaches may spread spores of fungi, and bacteria, which may get a hold in the intestine of any person eating contaminated food.

Mechanical spread of disease

This term is used, in contrast to contagion, to cover those insects which suck blood and feed restlessly, moving from one victim to another before being satisfied. These insects are liable to carry any parasites of the blood from one host to another, just as if one pricked a needle into the bloodstream of first one vertebrate animal and then another.

This method of infection is not clearly distinguished at the one extreme from accidental contagion, and at the other from the cyclical transmission to be discussed later.

Perhaps the most clearly defined disease in this category is *surra*, a disease of horses, camels and domestic animals such as buffalo in countries from those fringing the Sahara to the Phillipine Islands. It is caused by *Trypanosoma evansi*, and the related disease *mal-de-caderas* is caused by *T. equiperdum*. Both organisms are protozoa of the same genus as the organisms that in Africa produce sleeping-sickness in Man, and *nagana*, or cattle sickness in cattle and horses. The trypanosomes of sleeping-sickness and of nagana follow a cycle of development, part of which is passed within the body of a Tsetse-fly (see below). The uncommitted trypanosomes of surra and of mal-de-caderas are biologically more primitive, and it is considered that the ancestral trypanosome was probably dependent on mechanical transmission for its own survival, enabling it to survive in a new host when its previous victim died. Only later did species of *Trypanosoma* evolve that became dependent on a specific insect vector.

Among other diseases of domestic animals that may be carried mechanically by biting flies are anthrax, anaplasmosis and the bacterial disease tularaemia, but these can also be spread by accidental contagion and so they bridge the gap between contagious and mechanical transmission.

On the other hand *plague*, the ancient and virulent disease of rats that is passed on to Man by the bites of the Tropical Rat Flea, *Xenopsylla cheopsis*, lies somewhere between mechanical and cyclical transmission. The plague bacilli multiply so enormously in the intestine of the flea that they cause a blockage. The flea, unable to satisfy its hunger, bites repeatedly, perhaps on a number of different victims, and at each bite regurgitates some of the blood from its blocked gullet. As a result bacilli are injected into the blood of each victim of the flea's bites, which may be a rat one time and a man the next.

The spread of plague is thus mechanical in the sense that the parasitic organism is not compelled to spend any particular part of its life in the insect vector; yet the disease is cyclical in the sense that some lapse of time is necessary before an infected flea becomes congested enough to be a dangerous vector of plague.

Perhaps the most virulent of human diseases that are spread

mechanically, or by insect assisted contagion are those associated with the Human Louse. The rickettsial diseases – *typhus*, *trench fever* and *scrub typhus* (also known as *mite typhus* and *tsutsugamushi disease*) – are all closely related bacterial organisms which can multiply with great rapidity in human blood, with serious or fatal results. Typhus is caused by *Rickettsia prowazeki*, and trench fever by *Rickettsia quintana*, both of which are picked up by lice in the course of sucking human blood. They do not undergo any essential change when they are within the insect, but merely multiply in numbers. Indeed the first four may multiply so much inside the louse that it may be killed, and in this respect they are imperfect parasites.

The *Rickettsias* of the louse do not move into the salivary glands, and so are not passed on to a new human victim in the process of biting him, as is the case in malaria and similar cyclical diseases. Another man becomes infected by contagion, through the faeces of the louse, or its crushed body being rubbed into abrasions of the skin. Hence these diseases flourish under conditions where men wear the same clothing for long periods without changing it, and get dirty abrasions of the skin. Armies engaged in trench warfare, or long campaigns in the jungle, are prone to this sort of infection, as are civilian populations displaced by their activities.

Cyclical transmission of disease

This is the final phase of biological adaptation of a parasitic disease organism, by which one part of its cycle of development takes place in the blood of a vertebrate animal, and another, different part of its life-cycle requires it to find itself in the body of an insect or other vector.

Probably all such parasites arose from ancestors of a more general habit, as has already been noted in the trypanosomes. Mechanical transmission of blood diseases can occur only if the insect bites a second or a third victim almost immediately after the first. This is possible if the vertebrate hosts are crowded together in a herd, and are restless, disturbing the biting flies all the time with head or tail. Such mechanical transmission becomes less effective or impossible, when grazing animals are widely scattered, and when they have become tolerant of bites, and so do not drive the attacker away to seek another victim.

There would seem to be a biological advantage if the blood parasite can adapt itself to survive in the intestine of the biting insect, and so arrive at the first stage seen above in the typhus organisms of the louse. But these, too, depend in the end on chance, and a firmer basis for evolutionary adaptation occurs when the parasite not only changes its physiology and even its shape while living in the insect vector, but also its behaviour, migrating into the head so that it can find its way into a future victim of the biting insect.

Bloodsucking insects inject saliva before they suck up blood, mainly as an anti-coagulent, without which the proboscis would quickly become clogged and useless like a dirty fountain pen. A parasite that can make its way to the salivary glands and survive there, will therefore be injected into the next victim, and then find itself with a large reservoir of new blood in which to feed and multiply.

A disease-cycle of this type can flourish only if the host-animal and the vector-insect exist in the same area, at the same time, and even then only if the habits of the two match each other. For example a diurnal mosquito could hardly carry the disease organisms that live in nocturnal rodents, unless its habits were such that it frequented the places where the rodents sleep by day. In reverse this is the reason why a sleeping man is exposed to the bites of nocturnal mosquitoes, and must protect himself with a mosquito-net or prophyllactics.

Some blood parasites, such as the protozoa of *malaria* and of *sleeping-sickness* have an extended and complicated life-cycle. The asexual cycle of a malaria parasite produces a new crop of young parasites called merozoites in the blood of an infected person every few days, giving rise to the three main types of malaria: *tertian malaria* caused by *Plasmodium vivax; quartan malaria* caused by *P. malariae*, and *malignant* or *pernicious malaria* caused by *P. falciparum*. The attacks of fever in the sufferer come when a new batch of merozoites is released into the blood.

These asexual forms may be swallowed accidentally by a biting mosquito, and undergo no further development in the insect. But every so often there arise gametocytes, or sexual forms of the *Plasmodium* and it is these which continue their development cycle within the mosquito. Within about a fortnight they have undergone a sexual reproduction which produces a crop of active sporozoites, which are able to migrate to the salivary glands of the mosquito, thence to pass on to infect the next victim.

This type of disease-cycle, with its defined periods of development within the vertebrate host and in the insect vector, is obviously more difficult to coordinate between the three contributors: animal reservoir host, insect vector and parasite. So it might be expected that successful diseases of this degree of complexity are comparatively few. There are only a handful of trypanosomes carried in this way, and three major forms of human malaria, which are carried by mosquitoes of the genus *Anopheles*, though about twenty-five different species of this genus are important.

In contrast, certain virus diseases, of which *yellow fever*, *dengue* and the various forms of *encephalitis* are the most important, depend only on multiplication of the parasite. The adaptation and mutual adjustment that have taken place between the three contributors to such a cycle are perhaps less difficult. It would seem so, because such infections are carried by an assortment of mosquitoes of the tribe Culicini. Certain other parasites of the Phylum Nematoda, the so-called *filarias*, are carried both by anopheline mosquitoes and by culicines.

Bruce-Chwatt writes of the malaria parasites that they 'have probably evolved from free-living unicellular organisms, so ancient that none of their kind have survived the evolutionary past. It is probable that the parasitic mode of life inside the red blood cell was the key to the survival of the primitive parasites'.

Primitive members survive still in the other great group of parasitic protozoa, the trypanosomes, and are transmitted mechanically by Horse-flies and by *Stomoxys*, the Stable-fly. In particular this is true of *T. evansi*, the organism of surra. In Africa *Trypanosoma gambiense* causing human *sleeping-sickness*, and *T. rhodesiense*, causing the animal trypanosomiasis known as *nagana*, are both carried cyclically by Tsetse-flies of the genus *Glossina: G. palpalis* and *G. tachinoides* are the principal vectors of the former, and *G. swynnertoni*, *G. pallidipes* and *G. morsitans* of the latter. The complete epidemiology of the trypansomes and their vectors is not by any means fully understood, and it is thought that adaptive modification of the parasite has gone on even during the period that it has been studied, as the distribution and density of Man and cattle and game animals in Africa has changed considerably during the last hundred years.

South American trypanosomiasis

In South America *Trypanosoma cruzi* and *T. rangeli*, though generically related to the Old World trypanosomes, has obviously followed quite a different evolutionary history, and its insect vectors are bloodsucking hemipterous bugs of the genera *Triatoma*, *Rhodnius* and *Panstrongylus*.

The trypanosomes undergo cyclical development in the bug but, at least in the case of *Trypanosoma cruzi*, this has not yet progressed to the stage where the parasite is able to migrate to the salivary glands, and so be passed on during a subsequent act of biting. The infective stage of the trypanosome is passed out with the faeces of the bug and must gain access to another vertebrate host through a wound or an abrasion of the skin. This is only one stage further evolved than is the condition in lice and fleas, where the parasite merely multiplies within the insect host without developing into an 'infective form' that is visibly different. A second South American parasite, *T. rangeli* has taken the further step of migrating at least as far as the salivary glands and thus is able to penetrate a new mammalian host through the skin.

Trypanosoma cruzi has a reservoir in a wide variety of small mammals and birds, and all Reduviid bugs of the sub-family Triatominae pick up the trypanosome and cause it to develop into the so-called metacyclic infective form. When transmitted to Man, *T. cruzi* causes *Chagas' Disease*, so called because Chagas in Brazil in 1909 first showed that the disease was transmitted by the bug. The infection of Man with Chagas' Disease is in a sense accidental, and occurs only if he lived under conditions where the bug feeding on rodents and other small mammals can bite him. Biting takes place mainly at night.

Attacks on domestic animals

Insects attack other mammals in much the same way as they attack Man, and with similar harmful effects. Wild mammals may develop – or evolve by selection – a certain immunity to the blood-parasites, thus acquiring a tolerance which allows them to act as a reservoir of the disease without showing any symptoms. This is so of *Trypanosoma rhodesiense* in the blood of many of the big game animals of Africa, which themselves become a permanent reservoir from which *Glossina morsitans* can transmit the trypanosomes to horses, cattle and most other kinds of domestic animals. Wild animals may also become immune to the bites of insects, though they never become entirely indifferent to the nuisance of their attacks. Indeed in some badly fly-infested areas of north-east Africa the game move away at certain seasons when the flies are especially numerous.

Domestic animals are kept in an artificial environment. They are either herded together like sheep or cows, and kept under restraint by fences, or are forced to work in the open like horses, at a time when they would naturally seek shelter from the flies.

The various larvae that invade the human skin may also burrow into the skin of domestic animals. But the most serious are produced by organisms similar to, but specifically different from, those of Man. Thus human onchocerciasis is matched among animals by fistulous withers and poll evil in horses and blue-tongue in sheep. The virus diseases known as encephalitides are carried both to Man and to horses by certain Culicine mosquitoes, and it is thought that Man's encephalitis usually comes from an animal reservoir.

Three of the most troublesome pests of domesticated animals are the Sheep Blow-flies – *Lucilia sericata* and *L. cuprina* – and the Cattle Warbles of the genus *Hypoderma*. These last are highly specific to domestic animals and have no counterpart in Man.

Fleas heavily infest cats and dogs, pigs and poultry, and cause a certain amount of irritation to them, but the fleas of domestic animals are chiefly troublesome to the animals' owner. All such fleas will bite Man, though they breed only in association with their own host. Thus the source of much domestic infestation with fleas can be traced to domestic animals, but equally annoying infestations can arise from the Hedgehog Flea, *Archaeopsylla erinacei*, and from the so-called Chicken Flea, *Ceratophyllus gallinae*, originating from abandoned nests of small birds. The pig is exceptional among domestic animals in sharing the Human Flea, *Pulex irritans*, and so an infestation that originates in the pig-sty may establish itself indoors if enough undisturbed debris exists for the flea-larvae to feed in.

20
Insects and human food

Stored products

The habit of storing food, though by no means confined to Man, is one of his most distinctive characteristics. The food stored is most often of vegetable origin, flour, grain, dried peas, beans and similar seeds being by far the commonest items. Food of animal origin is not so easily stored unless it is either refrigerated – when it is too cold for insect attack – or preserved by desiccation, salting, treatment with a protective substance or canning. Hams and bacon are the principal dry goods of this type, and the problems relating to their preservation from insect attack are similar to those confronting dealers in hides, skins and furs. Certain insects are positively

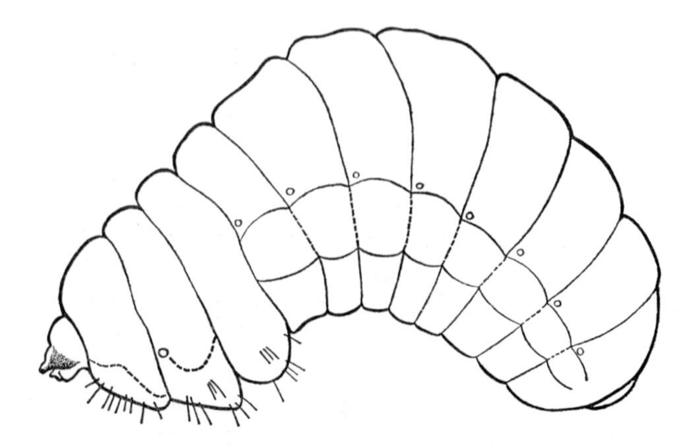

Fig. 71. Larva of a bean weevil (order Coleoptera), adapted to living in seclusion among an abundance of food (after Peterson).

attracted by such seemingly unpalatable food, notably Dermestid beetles, clothes moths, and cheese skippers (larvae of flies of the family Piophilidae).

Stored grains and farinose foodstuffs are attacked by a large variety of different species of insects, but these are nearly all the larvae and adults of beetles, and the larvae of moths. The attacks are non-specific in the sense that it is the physical conditions rather than the exact nature of the foodstuffs that provokes attack. Food in store offers to an insect a virtually unlimited amount, without the insect having to search for it. By the nature of things, food in store is normally left undisturbed for a long time, and so presents the insect with a secluded place with a suitable microclimate in which to breed through a succession of generations.

Attacks on growing crops

A growing crop is just as artificial a concentration as a store of food. Under natural vegetation it is rare to come upon a pure stand of any plant, and usually the individuals of each species are scattered among a great variety of others. Insects then have to search for their preferred food-plant in order to feed on it themselves, or to lay their eggs upon it, so that the newly hatched larvae will find suitable food at hand when they emerge.

When a field is cultivated the ground is ploughed, all the natural cover-plants being turned under and buried, then seeds are sown of one particular plant. These are either broadcast thinly over the whole area, or else are neatly sown in rows. When they sprout, the natural cover of grasses and broad-leaved plants is replaced by rows of little seedlings. All the plant-feeding insects are thereby concentrated on the crop-plants, and those which find this species palatable soon kill many or most of the young plants.

All parts of the plant, and all stages of its growth, are vulnerable to some insect or another. The larvae of a great many holometabolous insects, notably beetles, flies and some moths, live in the soil and feed on vegetable matter, including the roots of growing plants. This does not matter a great deal if the plant-cover is extensive, although even then if the insects are concentrated into a small area they may check growth and cause a brown area. Turf is particularly subject to this effect, because the leaves of the grass plants are kept mown short, and the normal well-being of the plant depends on its keeping an active root-system. Wireworms (the

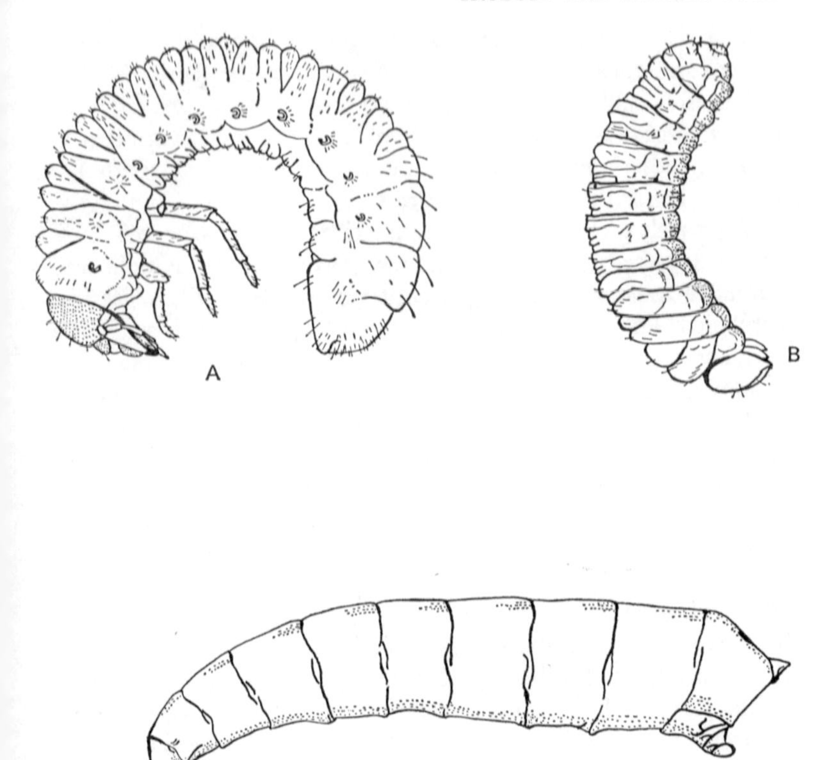

Figure 72. Three types of insect larva adapted to living among an abundance of food. A, a beetle, has retained a fully developed head and thoracic legs, but has sacrificed some mobility to enlarge the capacity of the abdomen; B, a honey-bee has become almost immobile, having lost external head-structure as well as legs; C, a fly maggot, has no legs nor external head-structure, but has developed 'mouth-hooks', and spiny swellings or 'pseudopods' along the body, becoming a highly mobile larva capable of feeding on a variety of foodstuffs.

larvae of Elaterid beetles), and leather-jackets (the larvae of Tipulid-flies, or daddy-long-legs) are notorious pests of turf.

The seedlings of cereals and root-crops – oats, wheat and barley on the one hand and turnips, swedes, mangolds, sugar-beet on the other – are vulnerable to root-feeders for the first period of their growth. After a time the root-system has become so much developed that destruction of part of it by insects is insignificant, and may even be beneficial.

The very young shoots above ground are attacked by leather-jackets and cutworms (the larvae of moths of the family Phalaenidae, often called Noctuidae). Besides feeding below ground level, these two groups of larvae come out on to the surface at night in moist weather and eat off the tender seedlings level with the ground, hence the name 'cut-worms'.

Vigorously growing, tender plants, with plenty of sap, are attractive to sap-feeding insects, and most particularly to nymphs and adults of most families of Homoptera: aphids, psyllids, jassids, cercopids. These insects can breed parthenogenetically through the summer season, and thus multiply at an astronomical rate. For example, Johnson [52] estimates that a very heavy infestation of *Aphis fabae* may discharge from one to ten million aphids into the air *each day* from one acre of beans. This must be considered in relation to the very large number of other aphids at the same period that are wingless and therefore incapable of joining in these flights.

Plant viruses

The great practical importance of aphids, however, arises not from their numbers and nuisance value, nor even from the direct effect on the plant of the loss of so much sap, but from the diseases that are transmitted by this means. There are many bacterial and fungoid diseases of plants that may be carried by any plant-feeding insect, or which can enter by contagion the wounds made by the insect in the plant. The sap-sucking insects of the various families of Homoptera, and particularly the aphids, have a close biological association with the plant viruses which live and multiply in the sap, and which depend for their own distribution if not their entire survival, upon being carried from one plant to another by sap-sucking insects.

The association between a particular virus and a particular insect or group of insects is close enough to suggest that this is a result of mutual evolution; that in fact the association corresponds more nearly to that of a human virus disease such as yellow fever than to one that is mechanically transmitted like tularaemia. A number of papers by L. M. Black make it clear that in general one virus depends on one method of transmission, and has evolved its own life-cycle and periodicity to suit the biology of the vector insect.

There is obviously a parallel between the transmission of disease among plants by sap sucking insects and the transmission of diseases among vertebrate animals by those insects that suck blood. Apart from the very few hemimetabolous insects that suck blood, notably the sucking lice and the Triatomid bugs, the sucking proboscis that has been evolved by hemimetabolous insects is used for piercing plants, and parasitic, disease-causing organisms have taken advantage of this in their evolution; whereas the sucking proboscis of Holometabola, especially well developed in Diptera, although deceptively similar to that of Hemimetabola, is used for piercing the skin of vertebrates and so transmits their blood parasites. As more detailed studies are made among aphids, the parallels between virus transmission among plants and virus transmission among vertebrate animals appear the closer.

Thus at one time mechanical, external, or non-persistent transmission was contrasted with non-mechanical, internal or persistent transmission. In the first a virus was spread only if the aphid fed on a second plant while its stylets were still wet with infected sap from the first plant; in the second type the virus passed into the body of the insect and there multiplied. The insect was then not immediately infective to a new plant, but when a sufficient colony of the virus had developed then the insect remained an effective vector of the virus for a considerable time. That is why the second type of transmission was called 'persistent'.

A point of importance in insects that develop as rapidly as do aphids is that the external infection of virus is lost when the insect moults, but the internal infection may be carried over into the next instar.

A great deal of work goes on in studying the details of particular diseases, because the best hope of controlling insect-borne diseases of plants is to try to break this close association between the virus and the vector. For instance a strain of plant that reaches its most juicy and attractive stage a little earlier or later in the season than others may miss the time when aphids are most abundant. Unfortunately, a rapidly reproducing insect such as a species of *Aphis* can quickly adapt its habit by natural selection. If food is available to it two weeks earlier than normal, this favours individuals that are about early, and so in a year or two's time the whole surviving population may have changed its habits accordingly.

Defoliation of established crops

When the individual plants of a crop are young and small they can easily be killed by a comparatively small density of insects. A piece of matted grassland, for example, may contain half a million wire-worms to the acre, and yet one would have to sift through one hundredweight of soil to find twenty of them. Such a population, feeding on the original turf, would make no visible impression, but if the turf is ploughed in and a few young plants of oats or barley are sown in neat rows, these same few wireworms will soon make a wreck of the crop.

The young plants soon grow to a size at which their roots are big enough to survive wireworm attack. The same is true of the leaves of the plant. The Turnip Flea Beetle, *Phyllotreta nemorum*, is a Chrysomelid beetle, both larvae and adults of which bite and chew the leaves of various cruciferous plants, and are especially destructive to young turnips and related crops. Though the beetle can be fairly numerous it is the vulnerability of the young plants that makes this insect pest particularly dangerous to them.

Once standing crops or trees and shrubs have begun to grow, they rapidly grow away from the pests of their infancy, and can take a great deal of superficial damage without being killed. Indeed a certain amount of such natural pruning could be beneficial. Moreover as the plant grows older it becomes less succulent, and the insects feeding on it may move away to another food-plant. The Chinch Bug, *Blissus leucopterus*, a major pest in North America, is a heteropterous insect of the family Lygaeidae, which sucks the sap of spring crops from April onwards. When the early cereals have ceased to be juicy enough for the bugs these move to corn (maize), which provides them with an extremely succulent crop in high summer. This bug is therefore a continuous pest throughout the growing season.

So the pests of established crops have a different relationship with the plant from that of the pests of young plants. They may be dangerous in comparatively small numbers if they attack some critical part of the plant. Examples of this are the Frit-fly of oats (*Oscinella frit*) and the Gout-fly of barley (*Chlorops taeniopus*), which weaken the stems of cereals so that an entire head may be sacrificed to the feeding of one larva. In this category come all those insects that feed in and damage the reproductive parts of a plant, when the flowers, fruit or seeds are the valuable part of the crop. Cotton is a major example, and has several insect enemies, in combating which a lot of money has to be spent.

The 'cotton stainers', *Dysdercus*, heteropterous bugs of the family Pyrrhocoridae, pierce the cotton boll and stain the fibre, thus making it commercially unusable. The cotton boll weevil, *Anthonomus grandis*, feeds as an adult in the seed pod of the cotton plant and lays eggs in it, so that the larvae complete its destruction. Since there are at least five generations per year this can be a particularly destructive pest.

The cotton bollworm, though similar in name, is a totally different insect: *Heliothis armigera*, a moth of the family Noctuidae (Phalaenidae). The caterpillars of this moth attack many other fruits and seeds besides cotton, and are particularly destructive to corn-cobs, eating a cavity in the grain.

Invasions of insects

The really devastating insects, however, are those which descend on a crop in enormous numbers and overwhelm it. No annual crop plant can survive being completely defoliated in the growing season. Growth is at least interrupted for the rest of the season, if the plant is not killed altogether. The sudden attack of a large number of insects, whether they arise in large numbers locally, or arrive in a swarm from elsewhere, is a visitation that has the attributes of an act of God. Among the most notorious of these are swarms of locusts.

Swarms of locusts

The sudden descent of a swarm of locusts has been a familiar experience in certain countries of Africa and the Middle East since earliest times.

Locusts are grasshoppers of a few species that not only multiply in great numbers, but under certain circumstances can become concentrated to an exceptional density of individuals and move about in a body usually known as a *swarm*. The word 'swarm' is often misused in entomology. Insects may be brought together because they all do the same thing at the same time: e.g. flies congregating in a roof-space. Yet they are not a swarm in the true sense unless they have some effect on each other, so that an individual behaved differently in the mass from its behaviour when solitary.

This is particularly true of locusts. Not only do they actively coalesce in a swarm, but both their physical appearance and their flight behaviour alter when they are gathered together. This change was elucidated by B. P. Uvarov in the *Phase Theory of Locusts*, and has been frequently confirmed and more subtly interpreted by both observation and experiment during the last forty years.

Individuals of the Desert Locust (*Schistocerca gregaria*) are predominantly green, and look rather like large grasshoppers as long as they live solitary, individual lives, as do the great majority of grasshoppers all over the world. If the locusts are congregated closely together, either as 'hoppers' (nymphs) or as winged adults, so that they see and touch each other, and stimulate each other into heightened physical activity, they also change in colour and pattern and become yellow or orange and very gaudy in appearance.

There is a subtle suitability about this colour change, the soothing green of the solitary and the strident, exciting pattern of the gregarious, and it may well be that the more vivid pattern of the gregarious phase increases the flicker stimulation of the eyes, and adds in this way to the general level of excitement.

The *solitaria*←→*gregaria* change is completely reversible, in particular among hoppers. Hoppers from a *gregarious* swarm that are isolated, either experimentally or by chance in the field, begin to revert to phase *solitaria* within a few days; whereas *solitaria* nymphs that are brought together change colour and show *gregarious* behaviour within a few hours.

The gregarious phase is closely associated with restlessness of the individuals and with *migration* of the swarm. Indeed, the migration is the crucial stage, because even today in the areas of the Middle East and Africa that are the principal sufferers from locust attacks, only a tiny proportion of the land is under cultivation of crops. Breeding-grounds of the locusts are mostly remote, isolated, swampy areas, and if the locusts stayed there they could become as numerous as they liked and the only sufferers would be themselves.

The menace of locusts is that massing of large numbers brings not only colour changes, but also a migratory urge, and the swarm flies long distances across vast areas of land before settling down again. Thus one swarm in 1954 in East Africa was observed over a flight lasting nine days, and covering 350 km. (about 220 miles). An FAO report of 1956 states that: '. . . swarms did not often breed within 1,000 km. of the areas within which they were produced, and there was nothing unusual in displacements of 1,000–

3,000 km. between breeding areas of successive generations. In three months in 1952 swarms from Ethiopia and Somalia reached India and Pakistan, 3,000 miles away.'

The winged adult locusts fly by day, and alight to feed every evening. It is doubtful whether they can to any extent 'choose' where they alight, in the sense of picking on any particular crop. They simply arrive out of the blue, settle on vegetation of some kind, and eat as much as they can. These swarms darken the sky and completely cover the vegetation on which they settle. It has been estimated that a large swarm of locusts may weigh about 15,000 tons, and each day eat as much food as one and a half million men.

Of course the very concentration of a locust swarm means that damage is confined to a limited area, and that one farmer may lose all his crops while another goes free. Moreover only a part of the vegetation eaten is cultivated crops. Yet the erratic nature of the attack, the completeness of the destruction, and its incidence in countries where the population, though relatively small, is concentrated, mean that until recent times the descent of a swarm of locusts might mean famine in the area.

In the major locust area of the Old World three principal species are concerned: the Desert Locust (*Schistocerca gregaria*); the African Migratory Locust (*Locusta migratoria migratorioides*), and the Red Locust (*Nomadacris septemfasciata*). Other sub-species of *Locusta migratoria* occur in the Middle East and Southern Asia, in Russia, in the Orient and in Madagascar. The Brown Locust (*Locustana pardalina*) originates from the arid Karroo area of South Africa, and the Moroccan Locust (*Dociostaurus maroccanus*) occurs from the Canary Island and Morocco, throughout much of the Mediterranean countries and as far east as Afghanistan.

In Australia the Australian Plague Locust (*Chortoicetes terminifera*) and the Australian Plague Grasshopper (*Austroicetes cruciata*) break out in troublesome swarms, and have become more significant as a result of an occupation and cultivation of this continent that are comparatively recent. In contrast the Rocky Mountain Locusts of North America (*Melanoplus* species), have declined since the prairies were cultivated in the last hundred years.

The first three species mentioned above are sufficiently important in themselves to call for a large international organisation whose activities are based on the work of the Anti-Locust Research Centre in London. The Migratory and Red Locusts differ from the

Desert Locust in that their breeding is restricted to certain small areas, mostly now discovered, and which are known as *outbreak areas.* Both study and control are simplified by the fact that they can be concentrated on these areas, and it is even possible to think that some at least of these areas may ultimately be changed so much that locusts no longer breed there.

Besides limited breeding areas, these two locusts also have *outbreak years,* and by careful observation of the known outbreak areas, it is often possible to forecast outbreaks, and even to prevent or minimise them by prompt attacks on the hopper bands from which they arise. For more than thirty years the outbreak areas of the Red Locust have been kept under surveillance, and increasing success has been achieved in killing the hoppers down to a limit at which no swarms escaped to migrate to other areas. Indeed the Red Locust Control Organisation which organises this work suffers from its own success. The Director of the Organisation wrote in 1960: 'Unfortunately practically nothing is known of the features of invasion (i.e. outbreak) areas that permit an incipient plague to burgeon', and so far it has been thought unsafe deliberately to allow a plague to arise in order to study its mechanism. This means that successful control measures remain rules of thumb, and that other outbreak areas could arise without anyone being able to recognise the risk in advance.

The Desert Locust has no restricted outbreak areas within which it can be controlled, and strictly speaking it has no plague years. The species range over the vast area from Southern Spain and Turkey to Tanganyika, and from the Gambia to India and Eastern Pakistan, an area divided between about sixty countries. Every year there are always swarms of the Desert Locust present somewhere in this area, and every one of these sixty countries has received major swarms at certain times, as well as having long periods without swarms.

The change from solitary to gregarious phase is brought about in this locust, as in others, by the mutual excitation of being crowded together, but the crowding arises differently from that of hoppers massed in a restricted outbreak area. Swarms of the Desert Locust build up by the accumulation of smaller bands of flying adult locusts which are brought together by the wind.

R. C. Rainey and others have demonstrated, by patient plotting of swarm movements and weather patterns over many years, that it is the wind that is principally responsible for the aggregations that lead on to gregarious behaviour. Photographs of swarms on a scale big enough to show the individual locusts indicate that although many of them face in the same direction as each other, and even move through the air for a time in that direction, there are many different components of the swarm milling about each other in different ways. These movements and the mutual reactions of the locusts involved in them, serve to keep the swarm together, and give it coherence, but the direction of movement of the swarm as a whole is down-wind. In some degree it is like a thunder-cloud, which has turbulence within itself, and some parts of which may be moving in every direction, but the cloud as a whole drifts inexorably with the wind.

The greatest and most devastating swarms of the Desert Locust, therefore, cumulate at points of converging winds, which are the low-pressure centres of atmospheric depressions. These are also areas of cloud-formation and rain. It seems that this is no coincidence, but has definite evolutionary significance. The area over which the Desert Locust occurs is predominantly arid, and the locust requires quite a lot of moisture for the maturation of its eggs in their pod in the ground. It has been estimated that at the time of laying the egg-pod contains only half the moisture that it will need, and the rest must be absorbed from the surrounding soil.

What happens, then is that locusts from a wide area are caught up in air movements and swept together; they become gregarious in behaviour, and therefore remain together in a coherent swarm, and the swarm travels with the wind, from north to south at some seasons and from west to east at others. This will bring the swarm to areas where rain is about to fall and where vegetation for overnight feeding is best. Eventually the locusts reach a major precipitation area and lay eggs in soil that will soon be moist enough for their requirements.

Thus by a process of selection over a very long period locusts arrive at places where they are most likely to succeed, but the link with weather is not exactly predictable. Swarms do not come every year to the same areas. The relation with weather is complex and the real picture is undoubtedly even more complex than is at present known. But the connection with weather has been established, and this makes it rather easier to warn particular countries that they are likely to receive a visitation from a large locust swarm.

The migration of insects

The movements of the Desert Locust provide one example of the general phenomenon of *migration* among insects, which may conveniently be discussed at this point because it is one of the ways in which insects intrude into human affairs. Migration is not a peculiarity confined to a few species of insects such as locusts and butterflies, but merely one extreme of mobility, contrasting with the extreme of sedentary life exemplified by the females of scale insects and Strepsiptera.

Most insects run or fly or jump short distances, in no particular direction, responding to the immediate stimuli of attraction or alarm. These various movements cancel each other, and the individual spends its life in a small area, much as people in country villages used to spend their lives within a radius of a few miles.

If for any reason at all the movements – that is, the displacements – of an insect are given a bias in one particular direction so that over a period of time it moves progressively further along a recognisable track, then we say that the insect is 'migrating'. It does not follow that all migrating insects do so for the same kind of reason, and it is certainly untrue to think that the migration is as purposeful as it appears to be. Even the migration of locusts into areas where rain falls and breeding is possible is an elusive kind of purpose, because the species could quite well survive in reasonable numbers in its original haunts, like the great majority of grasshoppers. It is as well to avoid comparison with the migrations of birds. Birds and insects are examples of entirely different organisation, and little enough is known about either to make comparisons other than confusing.

Migrating butterflies have a great attraction for students, but their chief importance to Man has been the occasional battle when they have flown in the face of one or other of the contending armies, and of course have been welcomed by the other side as a sign of divine support. Locusts are the perfect examples of the physical disturbance of the sudden appearance of very large numbers of insects, coupled with the devastation caused by the simultaneous feeding of so many.

Locusts are also large insects, and their sheer bulk makes an impact. Many tiny insects make sudden appearances in large numbers, an example being the 'Sun Pest', a Pentatomid bug, *Eurygaster integriceps*, which feeds on cereals in valleys in south-west

Asia. The bugs move into the mountains in hot weather, hibernate at slightly lower levels through the coldest months, and descend in great numbers again to the valleys to feed in the early spring, Williams [125] gives details of a large number of recorded cases. and discusses exhaustively the many problems arising from them. The reader is referred to his book for an authoritative study of migration in general, as well as to Southwood [100, 101].

At the present time much research is being carried on into the mechanism of migration, and into the stimuli by which it is brought about. The biggest factor in the operative mechanism is the *wind*. According to Hocking the maximum speed of flight of any living insect is about 35 m.p.h., and most flying insects move much more slowly than this. This means that the biggest and most powerful insects could just make headway against a wind of Force 5 to Force 7 on the Beaufort Scale, that is when the weather forecast says: 'wind fresh or strong'. Even these insects would have to make a concentrated effort, and would have to be kept heading in a constant direction by some powerful directional influence. The great majority of insects, including the butterflies that are the most conspicuous migrators, could not do this, nor could they withstand the buffeting that would be involved.

It seems, therefore, that migrating insects do not point themselves in a chosen direction and fly to a goal, as one is apt to imagine. Rather they launch themselves into the air and are passively carried down-wind regardless of their own orientation, and of any active movements that they may make in relation to the air immediately surrounding them. Thus both observation and photography of locust swarms show that individuals may be pointing in any direction. The individual movements of gregarious locusts serve the double purpose of exciting each other and maintaining them in gregarious phase, and also of giving the swarm coherence by turning the locusts into the swarm again when they reach its edge.

The question is often asked how far a House-fly can travel – a matter of some concern to Public Health Authorities and anyone who wants to blame a fly nuisance upon some fairly distant neighbour. It is impossible to give an exact answer because the displacement of the fly over the ground is a product of its own air movement and of the wind effect. A sustained flight in still air is probably restricted to a few hundred yards, and is seldom made by a fly under normal conditions. Most movements are short flights from one attractive smell to another, and keep the fly within the house or close by. If there is a steady wind, however, the fly continually drifts a little way down-wind, and if it once gets out of range of the powerful attractions of its normal habitat it may drift very far, in small stages, before settling into another locality. In this way a marked fly has been traced over a distance of about fifteen miles.

A spectacular nuisance arising from a 'migration' of this kind was that of the Seaweed-flies (*Coelopa frigida*) wandering from the south coast of England in the autumn of 1953. These flies have a normal movement of their own which might properly be called a migration, in that periodically they move in a compact band along the shoreline from one bank of decaying seaweed to another. This is an example of migration induced by overcrowding of the breeding and feeding medium. The direction of flight is governed partly visually, keeping parallel to the shore, and partly by the wind, being in the direction most nearly down-wind. In late 1953 there were persistently strong southerly winds, which therefore blew at right angles to the shoreline, and tended to force the flies out of visual contact with the shore.

Once contact with the shore had been lost the flies were in the same situation as the House-fly, and made aimless local flights, all the time with a down-wind trend. They must have been dispersed over a very large area of country, but were only detected when they came within range of the smell of trichlorethylene or a related compound, which acted as a powerful attractant and rallying force. Thus the flies swarmed into dry cleaners, garages, engineering works, and any place where such compounds were used as a degreasing agent or as a solvent.

Stimuli underlying migration

There are two schools of thought about the stimuli causing migration, the negative and the positive. The negative view emphasises the idea of *escape:* that a population of insects grown too big for its food supply launches itself into the air and migrates to somewhere else. This reaction undoubtedly occurs, as in the seaweed-flies just mentioned, but it suffers from the drawback that the insects may find themselves no better off after they have migrated. Indeed in the special circumstances detailed above the migration was fatal to the flies involved. This type of migration is a very haphazard affair, and must often be fatal. It recalls the suicide march of the lemmings.

It has been pointed out that the migrations of locusts bring them to areas where rain is falling, and where locust breeding is facilitated. This is represented as a positive advance (to the locust), but it is only so in respect of the huge populations generated by the gregarious habit. In fact it does not seem that any insect gains anything ultimately by *migration*, as distinct from mere *dispersal*. The latter, a scattering at random from a centre of origin, opens up many new habitats, and is undoubtedly one of the principal ways in which insects have successfully evolved. Wigglesworth has suggested that even the origin of wings can be traced to the evolutionary advantage of dispersal.

As far as human interests are concerned, the migrations of insects are always potentially a nuisance, if not a threat.

21 Insects in the home

Man is one of the animals that have a den, a nest or a regular habitation, and thereby provide a special environment that may be suitable for the feeding and breeding of various insects.

None of Man's domestic pests are peculiar to the house, and most of them live there in the same way in which they could be found living in birds' nests, in caves, or in the burrows of small mammals. In such situations they make use of shelter, protection from strong light, heat, rain or frost, but most of all they receive a steady supply of food from vegetable and animal debris, which will serve for the whole of their larval life.

It is because of this association with organic debris, that insects in the home are immediately thought of as being 'dirty', their presence an indication of insanitary conditions, and – even more important to most householders – a cause of shame towards one's friends. Besides the possibility of spreading disease, which has already been discussed, insects in the home are a nuisance by their presence, as well as causing damage to or loss of food, clothing, furniture, and even the structure of the house itself.

Insects as a nuisance—flies

Flies are the most obvious insect nuisance in the home. The House-fly, *Musca domestica*, is well named, because it may breed, and lives at all stages of its life in and around the home, and throughout the world a substantial majority of all individuals of this species live in close association with Man.

Musca domestica has a smooth maggot which feeds in organic debris. The larva requires a fairly narrow range of temperature, which is easily achieved by bacterial heating in a mass of decomposing organic matter. Breeding of the House-fly is thus encouraged by letting rubbish lie about in heaps which are undisturbed for at least two weeks. Dung of horses, pigs, and humans is specially attractive to the egg-laying female, but almost any kind of household refuse will do if it lies about and heats up to the desired extent, without drying out.

The larvae of the House-fly are seldom a nuisance in themselves, except psychologically, unless they infest wounds and dirty bandages, as they sometimes do. The adult House-fly creates the nuisance, by hanging about the house all day. The adult fly takes a great deal of liquid food and will suck at any wet surface. It feeds by regurgitating saliva and partly digested food from the crop, which softens and hydrolyses the substances on which it falls. The fly then sucks up all the available liquid, old and new.

Since the fly travels from faeces and other decaying materials to human food, and also settles on the person, and sucks sweat and moisture from eyes, lips or wounds, it is not only intensely irritating but also is a very potent agent in spreading many contagious diseases as well as intestinal infections. Wounded men after a battle have often suffered intolerably from the flies swarming round them.

No other fly is so ubiquitous and so pestaceous in the home as *Musca domestica*. The Lesser House-fly, *Fannia canicularis*, is the smaller species that is to be seen cruising round and round in the middle of the room, playing kiss-in-the-ring under a hanging lamp. This is only one species of a large genus, but the others are scarcely known except for one or two that breed in latrines. The larva of *Fannia* has fringed processes round the body, and lives in organic debris that is rich in nitrogen, particularly in places where there is urine. The droppings of birds, with their mixture of urine and faeces into guano, are very attractive to *Fannia canicularis*, which is abundant in chicken runs. Concentrated breeding of poultry in batteries or in deep litter houses has caused local invasions of neighbouring houses with these flies. A more purely domestic source of outbreaks of this fly is in woollen materials soaked in urine: a baby's cot mattress insufficiently protected by a rubber sheet; even a wooden tub used for storing the nappies while they were being accumulated for the large washing-machine. The adults of the Lesser House-fly may be irritating for their ceaseless activity, but they do not approach either food or the person in the manner of the House-fly.

Blow-flies, bluebottles (*Calliphora*), thrive around human habitations. Because they are big and buzz loudly, bluebottles create a nuisance out of all proportion to their numbers. One flying about a room can be quite maddening. The greenbottle (*Lucilia*) is smaller and quieter, and although often abundant around dustbins, does not come indoors so much.

The larvae of Blow-flies are more directly alarming than the adult flies. The maggots of *Calliphora* are notorious for turning up in 'blown' meat, but they have an equally troublesome habit of looking for a pupation site in many kinds of packages with which they have no obvious connection. After feeding, the larva crawls out of its food material and into a narrow space – this seems to be one of its urges at this time – and there pupates within the brown relic of the last larval skin, the puparium. Such pupae are often found in packages in warehouses, having come there from larvae that have bred in a dead rat, mouse or bird. A recent case is a warning against poisoning rats in the thatch of your country cottage, when you may get an invasion of large, juicy maggots of Blow-flies descending on you.

The larvae of *Lucilia* sometimes appear in very large numbers from the soil of the garden, usually close to the house, at a point where there are drains in the soil. At certain times of year, especially in late summer when heavy thunderstorms soak the soil and fill all the air-spaces that normally exist between the particles, the larvae come out on to the surface. They may be seen on a wet night, wriggling in hundreds on the concrete strip outside, or even coming under the door into the kitchen. In similar circumstances, invasions of leather-jackets – larvae of Tipulidae, or Crane-flies – may also come out of the soil.

Other invasions of flies, usually small, or even tiny, occur from time to time indoors. Some, like the tiny midges *Sciara*, breed in damp, mouldy places, and indicate a lack of ventilation in some corner of the house. *Drosophila*, a small fruit fly famed for the genetical experiments performed on it, occurs in most houses in small numbers, but if it suddenly becomes abundant there is probably a dish of overripe fruit, or a few kitchen scraps, forgotten in some corner.

Occasionally, too, a swarm – or more correctly a mass, or an aggregation – of flies comes in from outside. Such invasions of the loft by Cluster-flies (*Pollenia rudis*) and one or two other species of similar habit, or of the bedroom by the tiny yellow-and-black fly, *Thaumatomyia notata*, are seasonal, a winter sheltering of the adult flies. Those in the loft are little trouble as long as the weather remains cold, but during a mild spell, even in mid-winter, they may become active in a sluggish sort of way and cause a fly nuisance at a time when such things are not anticipated.

Mosquitoes have a special place as household pests. Most of the anophelines involved in human malaria feed indoors, and usually at night. Outdoor and diurnal feeders are less dangerous because their contacts with humans are fewer, and the mosquitoes are more easily seen and driven away. This is true, too, of many culicine mosquitoes, in particular of *Aedes aegypti*, the particular vector of urban yellow fever, which is predominantly a house-haunting species.

Besides coming indoors to feed, mosquitoes make great use of the shelter of houses to rest while the eggs are maturing, or during the heat of the day, or to hibernate through the winter of temperate countries. *Theobaldia annulata* and *Culex pipiens* are common indoors for the last reason. *Theobaldia* bites Man readily and painfully, but *Culex pipiens* feeds on the blood of birds, and so rates merely as a nuisance in the home. The related and almost indistinguishable *C. molestus* and *C. fatigans* do bite Man, and the former can be a great nuisance in cellars and other underground premises.

The house-haunting habits of mosquitoes make them vulnerable to the application of residual insecticides, and the spraying of the walls of houses has augmented or replaced the traditional oiling of pools as a convenient way of combating mosquitoes. It has the disadvantage inherent in all residual insecticides, that a resistant strain of the mosquito may be selected, or if not, then that another species of mosquito may move in to fill the niche left by killing off the first.

The true Sand-flies of the genus *Phlebotomus* like cool shade, and haunt the vicinity of houses in warm, arid countries. Besides being painful biters, Sand-flies carry a number of troublesome diseases.

Fleas

After flies one thinks of *fleas*. Fleas have specialised tastes for breeding, and their larvae live in the nest or habitation of a particular host animal, or range of related animals. At the same time adult fleas are quite catholic in their personal taste, and will bite wherever they can find blood to suck.

Fleas that breed in houses are the Human Flea, *Pulex irritans*, and the Cat and Dog Fleas, *Ctenocephalides felis* and *canis*. Their larvae live in crevices between floorboards, under skirtings, and in any undisturbed bedding material in kennels and sleeping boxes of the family pets. If the house is left empty for a few months the fleas that reach the adult stage remain within their cocoons until they are disturbed by the vibration of movement within the house. It is not uncommon, therefore, for new tenants to be greeted by an outbreak of fleas left behind as a legacy by their predecessors.

The Plague Flea, *Xenopsylla cheopis*, is a tropical Rat Flea. In temperate countries it is particularly associated with the Black Rat, and is fortunately of rare occurrence. Much more frequently encountered are the so-called Chicken Flea, *Ceratophyllus gallinae*, and the Hedgehog Flea, *Archaeopsylla erinacei*. The former is really a parasite of small birds such as nest in garden trees and shrubs. It is sometimes abundant in hen-houses. In the suburban garden this flea hops down from old, deserted nests into the long grass, and waits for some member of the family to brush past, when it jumps on to their legs. In a similar way visitors to the garden hut on a Sunday morning may pick up Hedgehog Fleas from a family of hedgehogs living under the floor. This nocturnal mammal is much commoner than is often realised, at least in Europe.

Cockroaches and silverfish

These rate mainly as nuisances, though they often cause alarm and despondency in respectable houses. Silverfish (order Thysanura) run about in warm places round the hearth or the kitchen stove, or appear from behind tiles round the kitchen sink – any place where tiny particles of food are available to them. They feed mostly at night, and get their common name from the flash of silver that one sees when the light is switched on and the *Lepisma* (silverfish) or *Thermobia* (firebrat) scurry away into hiding.

Cockroaches are universally detested, as much for their smell and their oily black-brown cuticle as for any real harm that they do. If they fall into soup they can be regarded as spoilers of food. If allowed to feed undetected they may seriously damage books, documents, clothing or leather goods, particularly if these are already old and rather frail. For instance on one occasion they attacked a segment of a 'big tree', *Metasequoia gigantea*, in a museum gallery.

Structural damage to buildings

Certain insects directly attack the structure of the house. Once again 'attack' is the wrong word, because this suggests that they single out structural timbers with a special vindictiveness. Like pests of crops, these insects merely follow their normal habits of feeding, but when they find themselves presented with an almost boundless supply of larval food successive generations have no incentive to wander away in search of any better place to live. The concentration of damage that results can be expensive and even irreparable.

Termites take pride of place here, because their activities are completely hidden, and may not be suspected until woodwork suddenly collapses. The larvae of various wood-boring beetles have a similarly disastrous effect on either structural timber or furniture. Powder-post Beetles (*Lyctus*) can reduce the leg of a chair to dust. Death Watch Beetles (*Xestobium rufovillosum*), and the recently increasing House Longhorn (*Hylotrupes bajulus*) have very large larvae which, by feeding over a long period, can severely weaken rafters and joists. They seldom cause sudden collapse, but by the time that they are discovered the timbers are often riddled beyond repair. Unless first-aid can be applied *in situ* by injecting some compound that will harden and give support to the riddled timbers, it is often necessary to dismantle and rebuild, say, a church roof, at very great expense.

The common Furniture Beetle, or woodworm, (*Anobium punctum*) is widespread, and is to be found somewhere in almost every house more than thirty years old. Its tiny holes disfigure furniture, floorboards and doorframes, and since the larvae have distinct preferences between one piece of wood and another – depending on the chemical differences arising from ageing, changing of the sugars and other sappy contents, and the action of moulds or fungi – the woodworm may riddle a small strip of wood so badly that it breaks under stress; e.g. a floorboard.

A basic difficulty about dealing with wood-boring beetles in the home is that they are not detected until the exit holes appear, and these are made by the adult beetles, which by that time have flown away to mate and start another generation of wood-borers. In certain places, notably the usual cupboard under the stairs, damage is cumulative because successive generations of beetles go on living there permanently, never leaving the house.

Lesser household pests are the Larder Beetles, especially *Dermestes lardarius*, which attacks anything that is dry and rich in protein. Others attacking anything from stored products to carpets are the larvae of Clothes Moths (*Tinea*) and the Brown House Moth (*Borkhausenia pseudospretena*). These are 'lesser' pests in the sense that the actual amount of damage is small, but its consequences may be troublesome, for example if a small moth-hole ruins an expensive garment or a carpet.

Insects in the mind

Most people never bother about insects unless insects bother them. Consequently few except entomologists have any idea how many insects exist, either as species or as individuals. It is surprising how a person noticing, say, a Tiger Beetle (*Cicindela*) for the first time will say he has never seen such a beautiful insect before, and can it possibly be British?

The average householder thinks of insects as vermin, and prides himself that his house is hygienic and free from them. Then one day he is bitten by a flea, or finds woodworm holes in a chair, or a few beetles in the jam in his larder. Looking around, he then finds to his horror that there are insects in every room, and calls in specialist advice, demanding fumigation and eradication of them all.

It is very difficult to convince such a person that a house without insects is as impossible as a house without germs. We have to live with them, as they have to live with us. If they are not a nuisance, and we are told that they are of a species that does no harm, the wise course is to forget about them (or, of course to study them!).

Unfortunately a few people find it impossible to forget the insects once they have been noticed. Beginning usually with an exaggerated fear of being bitten, it is easy to develop a general entomophobia, and to fear not so much actual pain as a sensation of being 'unclean'. This is an ancient and primitive fear, which in the past has often found religious expression, but which in the modern idiom is more often expressed in pseudo-scientific terms. Some people move house to get rid of fleas, and ironically they are just the people who often find another flea shortly afterwards in their new house. If there is any explanation of this other than mere coincidence it is possible that this particular person has an odour attractive to fleas: a dreadful idea to the person concerned, though understandable to the detached entomologist; for, after all, fleas are bloodsucking insects to whom smelling out a potential meal is a matter of life and death.

A few people of neurotic tendency get the idea that they are breeding or 'generating' insects in their skin, and consult a succession of both entomologists and doctors in the vain hope of finding relief. Once such an obsession has been developed there is little anyone else can do to help, because neither rational argument nor ridicule is usually effective. Fears such as these are made worse by commercial advertising of insect remedies, which naturally paint an alarming picture of the depredations of insects in the home. It is important to keep all insects in their correct perspective.

22
Beneficial insects

The insects are most often thought of as enemies of Man, but their beneficial activities should not be forgotten. Out of all the million or so species of insects, only a very few have been deliberately cultivated by Man. This is rather surprising in contrast with the very large number of plants that Man has cultivated either as commercial crops or for aesthetic attractiveness in gardens.

Man has made use of relatively few of the vertebrate animals, but even so, he has domesticated a useful selection of mammals. Insects have proved even less adaptable to human affairs, and the few that have been 'cultivated' are valuable not for themselves, but for their products.

Bees

The honey bee (*Apis mellifera*) is of course the outstanding example. Here Man has taken one species of a social insect, which was originally tropical and lived in colonies of indefinite duration, perpetuating themselves by sending out swarms to start new colonies. These colonies survived through unfavourable seasons, when flowering plants were not accessible, by storing food in the form of honey in cells in their colony.

Man long ago took advantage of this, first to rob the wild colonies as bears do, and then to transfer himself and his colonies of the bees to temperate countries, where their storing habits enable them to survive through the cold winters. By periodically robbing the honey-stores the bee-keeper made his labour worthwhile, especially in the days before he could get sweetness from sugar-cane and sugar-beet. Beeswax is another important product of the honey comb and is still used in great quantities in polishes, although lately somewhat replaced by synthetic substances.

So far has this process of cultivation of bees progressed that *Apis mellifera* is almost entirely a domesticated insect. Wild honey bees occur in the warmer countries, but those that are truly wild, and not feral – descended from swarms escaped from a domestic hive – are recognisably distinct from the hive-bee, as a variety if not as a distinct species such as the Indian *Apis dorsata*.

Silkworms

The silkworm, *Bombyx mori* is the other thoroughly domesticated insect. Here the product is silk, with which the larva spins a cocoon before pupating. A great many insects of all levels of evolution produce silk from a variety of different glands. Sometimes, as in silkworms, it comes from glands associated with the mouthparts, while other insects produce silk from cuticular glands, for example in the legs.

It is said that each cocoon of *Bombyx mori* produces a single strand of silk about one thousand feet long, and that over twenty-five thousand cocoons are required for each pound of silk. Silk-production is a highly specialised industry, particularly practised in China for many centuries, and nowadays carried on only on a small scale in southern Europe, England and North America. Though synthetic fibres have replaced silk in many of its uses, including many garments, silk fibre has a unique texture and quality that will probably mean that it will never go out of use entirely.

The silkworm moth has been taken over so completely by the silk industry, bred and selected into a number of strains as carefully as dogs or horses, that the original wild species has disappeared entirely from view. *Bombyx mori* is now only known as a domesticated insect.

Cochineal and Lac insects

The remaining 'cultivated' insects are a few Homoptera of the super-family Coccoidea, the Lac Insect (*Laccifer lacca*) and the Cochineal Insect (*Dactylopus coccus*). These are both *scale insects*, the females of which live a sedentary life on a plant, covered and protected by a resinous scale. In general scale insects are a pest, because their incrustations and their steady withdrawal of sap from the plant can be destructive. In these two examples, the first in the Orient and the second in Central America, the scale has been turned to human advantage as a product for which Man can find a use.

The incrustations of the lac insect are melted to form *shellac* and *gum-lac*, still much in demand in resins, varnishes and sealing compounds, while cochineal is used as a dye and a colouring matter in foods. Again, as with silk, synthetic substances threaten to replace the natural product, tediously collected from a very large number of insects.

Incidental benefit from insects

The biggest benefits that Man gets from insects, however, are incidental, not specially sought, and seldom appreciated, but in total far outweighing the rather meagre products of the bee, the silkworm, and the lac and cochineal insects.

Though individually insignificant, insects are astronomically abundant over the countryside, and they are an essential component of the life-cycle of almost all other animals and plants.

Some of the ways in which insects benefit Man are as follows:

Pollination of crops

The colour and shape of flowers attracts insects and thereby ensures pollination. If insects did not exist, and pollination was entirely by wind dispersal, plants would still have to have 'flowers' in the sense of sexual parts with exposed pollen-bearing organs; but they would never have evolved the shapes and colours and scents that attract insects.

We think of bees as the prime pollinators, and so they are if we include not only the honey bee but the very large number of wild solitary as well as social bees, including all the Humble Bees. Yet flies are probably as important in the aggregate. A clump of flowering shrubs in the summer will have more flies than bees on it. Other insects are presumably less important, because they do not flit so readily from flower to flower, but bugs and beetles also contribute.

Insects as food

Insects are not often eaten deliberately as human food, except among primitive wandering tribes such as the Australian Bushmen, the South African Hottentots, and some Indians of the Amazon Basin – people who live precariously on what they can find, without cultivating crops or tending herds. They find juicy beetle larvae, caterpillars or locusts. Some African tribes living beside the Great Lakes have at times collected the great swarms of midges that hang over the water, and pressed them into 'Kungu Cake', and North American Indians are on record as having collected the masses of oviposition females of flies of the genus *Atherix* which cling to branches over water.

The fact that insect-eating persists largely in tribes that have not advanced from early ways of life suggests that probably primitive man ate insects as a regular part of his diet, as chimpanzees and other Anthropoids still do. Besides being tasty morsels introducing variety into a dull diet, they provided a welcome course of animal protein, which became superfluous once Man had meat and milk from his own herds.

Lately there has revived a fad in certain parts of the world notably in the Orient and Mexico, for tinning juicy insects such as caterpillars or beetle-larvae. Even such unpromising subjects as ants are tinned and exported to more sophisticated countries as a 'delicacy'.

The important food value of insects, however, is indirect. They form part of elaborate food-chains in the nutrition of fishes and birds. For instance one swallow may catch and eat a great many insects in one day. The study of the pellets passed or regurgitated by birds shows what an important item insects are in their diet, and conversely how important a part the birds play in regulating the numbers of insects.

Control of other insects

Finally, of course, insects are important to us as a control for other insects. Predatory insects such as Dragon-flies and Robber-flies catch and eat many other flying insects, but their choice of prey is governed by what happens to be about, and so they operate as a control only in the most general way. This is true, too, of the carnivorous larvae, which generally eat anything they meet and catch. Parasitic insects are rather more specific, the Hymenoptera and Diptera providing most of the parasitic larvae which over the years exert a very close control over the population of perhaps most herbivorous insects.

The last is probably the most important beneficial activity of insects to Man but is too big a subject (see Bibliography).

23
Insects in the future

There is more to this topic than just a pleasant philosophic speculation. Up to the end of the nineteenth century, the evolution of animals and plants was a natural phenomenon: i.e. it went on its way without being affected to any significant extent by the conscious activities of Man. This was the evolution that was demonstrated by Darwin, and as so often happens, the existence of a phenomenon and its mechanisms were understood only at the very moment when everything changed.

Of course the fundamental principle of evolution remains true. Natural selection is an observed fact. The insects themselves give the most convincing additional proof of Darwin's reasoning by the alacrity with which they produce strains resistant against chemical attack. What has altered, and what has tilted the balance of evolution in new directions, is the enormous increase in the scale of Man's interference with nature.

Man first began to interfere with nature when he tamed and domesticated animals: the horse to ride, the dog for help in hunting, pigs, sheep, and cattle for the sake of their products. Nothing is more unnatural than gardening, and agriculture is a deliberate and studied interference with nature. Harvesting a crop has been described as an 'ecological catastrophe' (Chauvin, 1967).

During the historical period of Man's evolution, until quite recently, men have been too few, and their efforts altogether too puny, to have more than a very local effect on the natural environment. Exceptions have occurred in certain marginal areas where water was already short, and where little was needed to start a sequence leading to increasing aridity. The most notable example, is, of course, the marginal areas of the eastern Mediterranean, the 'cradle of [European] civilisation'. Perhaps the major factor was a drop in rainfall after the peak of the last ice age, about 17,000 years ago (Lamb, 1965), but the destructive agriculture of Man, and especially his herding of animals together into such numbers that they could be supported only by a nomadic existence, exhausting the vegetation as they went, contributed to a progressive aridity and erosion, leading to the deserts of the present day.

The existence of other large desert belts – Asiatic, Australian, and Central American – shows that Man alone is not entirely to blame. Perhaps the biggest example of a major ecological change brought about by Man alone is the transformation of the North American prairies. Here the last hundred years have seen the disappearance of the buffalo – and the Red Indian – the retreat of the American Locust (see Chapter 20) and in some areas the terminal problem of soil erosion, only partly arrested by expensive and elaborate irrigation schemes.

The effect of Man on the physical environment, like so many phenomena including his own 'explosive' population growth, follows an exponential law, the rate of increase itself increasing with time. In the past the rate of change, though always increasing, has remained so low that its results could be readily assimilated. A gradual change of the physical surroundings of insects was easily met by evolution and adaptation. An interesting paper by Perring (1965) on the advance and retreat of the British flora, stresses that in a mixed habitat such as is packed into the small area of the British Isles Man's interference has only a local or temporary effect. Although the landscape as we know it in the long-settled and heavily populated areas of western Europe is entirely man-made, this has happened so gradually and fragmentally that it is difficult to find many examples of direct effect on the insects. The draining of the English fens, and the eventual disappearance of malaria from this region through breaking of the cycle between Man and mosquito is one obvious example.

In future the scale of Man's activities is going to be very much bigger, and very much more rapid in execution. Air travel, massive motor transport, and even more massive earth-moving equipment, concentrated by radio's instant communication, mean that there is now no part of the earth's surface that is not liable to be demolished and rebuilt as a consequence of some decision made by someone in an office thousands of miles away. Nothing is physically impossible any more. It is true that nothing Man can do at present comes anywhere near disposing of the amount of energy contained in an Atlantic depression, a thousand miles across, and not finished in a fraction of a second, but blowing continuously for four or five days, perhaps for a week or more: but nuclear power can concentrate energy into a small area and a very short instant of time, and in this way literally move mountains.

Some ideas already put forward, though fantastic, are no more preposterous than talking of making a 'soft landing' on the moon, and even of coming back again. There are schemes to use nuclear power to remove, or at least breach, ranges of mountains, thus not only altering drainage over a big area, but, more important, making major changes in the circulation of the air, and so altering the pattern of global weather. Any action of this kind would present the plants, and therefore the insects, of huge areas with changes in a short space of time of a kind that previously required a geological era to come about.

Another idea is to use nuclear power to break up the polar ice-pack, and in this way hasten its demise. All human history has been lived in the shadow of the ice ages, and all the insects that Man has ever encountered have lived in the post-glacial period. The vegetational environment of insects and of Man is determined not only by the advance and retreat of the ice itself, but by consequent changes in rainfall and in underground water-levels over the rest of the world. The polar ice-cap is thought to be always in a state of unstable equilibrium, tending either to grow or to shrink, and that quite rapidly, much faster than most geophysical phenomena. The critical conditions are those round the margin, where precipitation either as snow or as rain, is greatest. The more snow falls, the more ice is formed, the more heat is reflected away uselessly (therefore less evaporation), and the more snow is likely to fall – dare one call it a snowball effect? Conversely, destruction of ice round the edges on a big scale might start this process in reverse, ultimately perhaps destroying the northern ice-cap completely. Some meteorologists believe that a pole with no ice might at last be in stable equilibrium and that the ice would not return. It is not stated whether a similar treatment would be given to the southern ice-cap, and what would happen if this were left untreated.

Such a globe would have quite different weather and vegetational belts from the present earth; it would also have a higher sea-level because of the additional water released, and a lower salinity because the ice-melt would be fresh.

It can be seen, therefore, that the environment facing the insects in future centuries will be very different from any they have had in the past, and, even more important, that these changes may come catastrophically quickly over bigger areas than ever before. What will the insects do? One thing is certain, that they will evolve. Their great advantage over most kinds of animals, particularly big mammals, and most birds, is their enormous fecundity. There is a poor future for any animals that reproduce slowly, because they will not be able to keep pace with their changing environment. 'Studies of recent mammal extinctions show that Man has been either directly or indirectly responsible for the disappearance or near-disappearance of more than 450 species of animals' (Newell, 1963).

It is seldom realised when talking about insects that the adult insect is the rare exception. Except among the social insects, where the larvae are specially protected, and where the system depends on the cooperation of very large numbers of adults, the adults are the few survivors from a large number of young larvae or nymphs. Which individuals are those survivors depends to some extent on mere chance, but even slight differences in the probability of being killed by adverse conditions, or eaten by some predator determine the course of evolution and adaptation to the environment of the time. Insects, therefore, can react swiftly to sudden changes of their environment simply by having a different set of individuals survive to become adult, and to pass on their characteristics to their descendants.

One undoubted effect of Man's interference with nature is to make the vegetation and other ecological factors more *uniform* than they were. A pure stand of any tree, herb or grass is rare in nature, and does not last long. A piece of land left without human intervention goes through a sequence of changes, and if the climatic conditions remain constant it reaches a climax vegetation suited to its location and climate. This is a process of natural selection among the plants and is accompanied by a similar selection among the insects. At any stage, even the climax, the vegetation provides a multitude of ecological niches for different kinds of insects. If an area is cleared by Man – trees felled, bushes rooted out, soil ploughed and sown with a uniform crop, then the insects able to live there are drastically reduced in variety. In a single crop there may be large numbers of insects, but few species. Therefore, since Man's activities always lead to a simpler, more uniform vegetation than unaided natural selection, these activities also favour fewer species of insect; but by the same process these are the species which have become adapted to the special conditions created by Man, and therefore the most likely to increase in numbers and become a pest.

It is fairly certain, therefore, that as time goes on the *variety* of insects will become less. Certain types of habitat will decline, notably swampy and boggy areas, and natural waters generally, as drainage is increased, both to make more land available for cultivation, and to fill reservoirs to supply the ever increasing demands for water for big cities. There may be a corresponding decline of Dragon-flies, May-flies, and the other groups with aquatic larvae, which have mostly persisted for a very long time in evolution. On the other hand there will be new opportunities for breeding in 'container habitats', small, temporary accumulations of water, in empty tins, old bath-tubs, cisterns, and the other debris that accumulates everywhere these days, and which is becoming increasingly difficult to dispose of. Some of the most dangerous mosquitoes breed in such places, and their lead will be followed perhaps by insects of other groups that will find Man obligingly making new habitats for them.

The insects that remain in man-made forests, and areas of uniform crops, will be heavily attacked by insecticides and chemical sterilants. This will help to kill off the miscellaneous insects, but by the nature of things, the insects that survive will be those inherently ablest to adapt themselves by rapid physiological selection. This is why the House-fly, *Musca domestica*, is such a successful insect: it exploits the conditions of human habitations, and its larvae will feed successfully in a greater variety of waste materials than any of its rivals.

In fact it is clear that insects of the future will not change their shapes as much as they did under the slow evolution of the past, but will make rapid physiological adaptations. It is sad to think that few really new shapes of insects will arise for future entomologists to collect, and that probably the outlook for the future is dull and drab; at least so long as Man himself sets the scene. Though Man's activities may produce exciting new environments, the insects will not have time to respond by structural evolution before all is swept away and altered into something else.

The insects that will flourish will be those that can exploit sources of food that will always be present in any foreseeable future, the sap of young plants, the fauna and flora of temporary water, and decaying and fermenting materials of all kinds. The first will ensure the future of aphids and other plant-feeding bugs. Insects breeding in container habitats, as we have said, will find plenty of scope. And in spite of the best hygienic ideas there will always be plenty of vegetable and animal remains to decay and ferment, so Blow-flies, House-flies and such yeast-loving flies as the Drosophilidae will produce a great number of almost indistinguishable species. That this process is going on already is shown by a recent work by D. Elmo Hardy in the series 'Insects of Hawaii'; Hardy lists two hundred species of *Drosophila* in the Hawaiian Islands, and estimates that double this number exist there. This is the genus of flies that is used for genetical work because it can not only be bred easily in the laboratory, but shows so much variation and mutation that it is a perfect experimental animal. It seems to have found an open-air laboratory in the Hawaiian Islands.

To separate and distinguish all these closely knit species of the future will call for a new type of systematist using a numerical taxonomy, and quantitative methods, perhaps entirely computerised. Entomology as a science will take new directions; it will be more like accountancy or banking, and will be very dull.

Appendix

TABLE 3

A LIST OF THE ORDERS AND SUB-ORDERS OF INSECTS

The figures given thus: 23/350, indicate first the number of species occurring in Britain, and second the world-figure. These are, of course, often mere guesses, but they give a rough indication of the comparative sizes of the orders.

APTERYGOTA				
Thysanura	23/350	2 families: Machilidae; Lepismatidae		
Diplura	11/400	3 families: Campodeidae; Japygidae; Projapygidae		
Protura	17/43	3 families: Eosentomidae; Prosentomidae; Acerentomidae		
Collembola	250/1500	10 families		
PTERYGOTA				
Palaeoptera				
Ephemeroptera	46/1500	3 super-families	Ephemeroidea	4 families
			Baetoidea	5 families
			Heptagenoidea	4 families
Odonata	42/5000	3 sub-orders	Zygoptera	16 families
			Anisozygoptera	1 family
			Anisoptera	6 families
NEOPTERA				
Hemimetabola/Exopterygota				
Plecoptera	34/1300	7 families		
Grylloblattodea	0/6	1 family	1 genus, 6 species.	
Orthoptera	/10,000	2 Sub-orders	Ensifera	
			Tettigonioidea	6 families
			Grylloidea	2 families

NEOPTERA—*cont.*

Orthoptera				
			Caelifera	
			Acridioidea	5 families
			Tridactyloidea	2 families
Phasmida	0/2000	3 families	Bacteriidae	
			Phyllidae	
			Phasmidae	
Dermaptera	10/900	3 sub-orders	Forficulina	6 families
			Arixeniina	Arixenidae
			Hemimerina	Hemimeridae
Embioptera	0/150	7 families		
Dictyoptera	34/5000	Sub-order	Blattaria	1 family
			Mantodea	1 family
Isoptera	0/1700	5 families	Mastotermitidae	
			Kalotermitidae	
			Hodotermitidae	
			Rhinotermitidae	
			Termitidae	
Zoraptera	0/16	1 family	Zorotypidae	
Psocoptera	70/1000	17 families		
Mallophaga	260/2600	3 sub-orders	Amblycera	3 families
			Ischnocera	2 families
			Rhynchopthirina	1 family
Siphunculata (Anoplura)	24/225	6 families		
Hemiptera				
Sub-order Homoptera		3 series		
		Coleorrhyncha		1 family
		Auchenorrhyncha		
		Cicadoidea		4 families
		Fulgoroidea		18 families
		Sternorrhyncha		
		Coccoidea		16 families
		Psylloidea		1 family
		Aleyrodoidea		1 family
		Aphidoidea		3 families
Sub-order Heteroptera		2 series	Cryptocerata	9 families
			Gymnocerata	41 families
Thysanoptera		2 sub-orders	Terebrantia	5 families
			Tubulifera	7 families
Holometabola/Endopterygota				
Neuroptera				
Sub-order Megaloptera		Super-family Sialoidea		2 families

		Super-family Raphidioidea		1 family
Sub-order Planipennia		Super-families	Ithonoidea	1 family
			Coniopterygioidea	1 family
			Hemerobioidea	1 family
			Myrmeleontoidea	5 families
Mecoptera	4/300	Sub-orders	Protomecoptera	2 families
			Eumecoptera	3 families
Trichoptera	188/3000			22 families
Zeugloptera	5/*c.* 100			1 family
Lepidoptera	2000/100,000			
Sub-order	Monotrysia			11 families
Sub-order	Ditrysia	Super-families	Hesperioidea	many families
			Papilionoidea	
			Tinaeoidea	
			Pyraloidea	
			Noctuoidea	
			Geometroidea	
			Sphingoidea	
			Bombycoidea	
			Tortricoidea	
			Calliduloidea	
			Castnioidea	
			Cossoidea	
			Psychoidea	
Diptera	5000/80,000			
Sub-order Nematocera				19 families
Sub-order Brachycera				14 families
Sub-order Cyclorrhapha		Series Aschiza		5 families
		Series Acalypterae		many families
		Series Calypterae		many families
		Series Pupipara		3 families
Siphonaptera	51/1000			
		Super-family	Pulicoidea	2 families
		Super-family	Ceratophylloidea	10 families
Hymenoptera	6000/100,000			
Sub-order Symphyta				14 families
Sub-order Apocrita				many families
super-families		Evanioidea		
		Ichneumonoidea		
		Cynipoidea		
		Chalcidoidea		
		Trigonaloidea		
		Proctotrupoidea		

NEOPTERA—*cont.*

Hymenoptera			
		Bethyloidea	
Super-family		Scolioidea	
		Pompiloidea	
		Sphecoidea	
		Vespoidea	
		Apoidea	
		Formicoidea	
Coleoptera	3700/220,000		
Sub-order Adephaga			9 families
Sub-order Archostemmata			2 families
Sub-order Polyphaga			many families
Series Haplogastra	Super-families	Scarabaeoidea	
		Hydrophiloidea	
		Staphylinoidea	
		Histeroidea	
Series Cryptogastra	Super-families	Byrrhoidea	
		Dascilloidea	
		Dryopoidea	
		Rhipiceroidea	
		Buprestoidea	
		Elateroidea	
		Cantharoidea	
		Dermestoidea	
		Bostrychoidea	
		Chrysomeloidea	
		Curculionoidea	
		Cleroidea	
		Lymexyloidea	
		Cucujoidea	
Strepsiptera	17/300		6 families

Bibliography

1 ALBRECHT, F. O. (1953) *The anatomy of the Migratory Locust.* London.

2 ALEXANDER, R. D. & BROWN, W. L. (1963) Mating Behaviour and the origin of insect wings. *Occas. Papers Mus. Zool. Univ. Michigan.* **628**:1–19.

3 ANDREWARTHA, H. G. (1962) *An introduction to the study of animal Populations.* Univ. Chicago. 281 pp.

4 BANRHARDT, C. S. (1961) The internal anatomy of the silverfish *Ctenolepisma campbelli* and *Lepisma saccharina. Ann. ent. Soc. Amer.* **54**:177–96.

5 BARRASS, R. (1964) *The locust: a guide for practical laboratory work.* 59 pp.

6 BAUMHOVER, A. H. et al. (1955) Screw-worm control through release of sterilised flies. *J. econ. Ent.* **48**:462–6.

7 BILIOTTI, E. (1966) The limits of biological control. *Pest Articles and News Summaries* (PANS). **12**:47–56.

8 BIRD, R. D. & ROMANOW, W. (1966) The effect of agricultural development on the grasshopper populations of the Red River Valley of Manitoba, Canada. *Canad. Ent.* **98**:487–507.

9 BORROR, D. J. & DELONG, D. M. (1954) *An introduction to the study of insects.* N.Y.: Rinehart. 1030 pp.

10 BRAIDWOOD, R. J. (1960) The agricultural revolution. *Scient. Amer.* **203**(3):131–48.

11 BRUES, C. T. (1946) *Insect dietary: an account of the food habits of insects.* Harvard Univ. Press. 466 pp. 68 figs. XXII pls.

12 BURNETT, G. F. (1965) Some newer ways of attacking tsetse flies. *J. Roy. Soc. Arts.* **114**:16–29.

13 BURR, MALCOLM (1954) *The insect legion.* London: Nisbet. 2nd Ed. 335 pp.

13a BUTLER, C. G. (1967) Insect pheromes. *Biol. Revs.* **42**: 42–87.

14 CAMERON, EWEN (1961) *The cockroach* (Periplaneta americana L.): *an introduction to entomology for students of science and medicine.* London: Heinemann. 111 pp. 31 figs.

15 CAMPION, D. G. (1966) The present status of research in chemosterilants in North and Central America. *Pest Articles and News Summaries* (PANS). **11**(A):467–91.

16 CARBONELL, O. S. (1959) The external anatomy of the South American semi-aquatic grasshopper *Marellia remipes* Uvarov (Acridiidae: Pauliniidae). *Smiths. misc. Coll.* **137**:61–97.

17 CARTHY, J. D. (1957) *Animal Navigation.* London: Allen & Unwin. 191 pp.

18 CARTHY, J. D. (1958) *An introduction to the behaviour of vertebrates.* London: Allen & Unwin.

19 CHAMBERLAIN, W. F. (1962) Chemical sterilization of the screwworm. *J. econ. Ent.* **55**:240–48.

20 CHAO, H. (1953) The external morphology of the dragonfly *Onychogomphus ardeus* Needham. *Smiths. misc. Coll.* **122**(6):1–56.

21 CHENG, TIEN—HSI (1963) Insect control in mainland China. *Science.* **140**:269–77.

22 CHINA, W. E. 1962. South American Peloridiidae (Hemiptera-Homoptera: Coleorrhyncha). *Trans. R. ent. Soc. Lond.* **114**:131–161.

23 CLARKE, J. F. G. *et al.* (1959) Studies in invertebrate morphology, published in honor of Dr Robert Evans Snodgrass on the occasion of his eighty-fourth birthday, 5 July 1959. *Smiths. misc. Coll.* **137**:416 pp. [includes TUXEN, S. L. on Apterygota].

24 CLAUSEN, C. P. (1941) *Entomophagous insects.* N.Y.: McGraw Hill.

25 COLHOUN, E. H. (1960) Acclimatization to cold in insects. *Ent. exp. et appl.* **3**:27–37.

26 CUMBER, R. A. (1949) Humble-bee parasites and commensals found within a thirty-mile radius of London. *Proc. R. ent. Soc. Lond.* (A)**24**:119–27.

27 CUMMINS, K. W., MILLER, L. D., SMITH, N. A., FOX, R. M. (1965) *Experimental Entomology.* N.Y.: Reinhold. 176 pp.

28 DALY, H. V. (1964) Skeleto-muscular morphogenesis of the thorax and wings of the honey bee *Apis mellifera. Univ. California Publ. ent.* **39**:77 pp.

29 DEMERECQ, M. [Ed.] (1950) *Biology of* Drosophila. N.Y. & London. 632 pp.

30 DETHIER, V. G. (1962) Chemoreceptor mechanisms in insects. *Soc. exp. Biol. Symp. XVI: Biological receptor mechanisms:* 180–96.

31 DU PORTE, E. M. (1959) *Manual of insect morphology.* N.Y.: Reinhold 224 pp.

32 EDWARDS, C. A. & HEATH, G. W. (1964) *The principles of agricultural entomology.* N.Y.: Reinhold. 416 pp.

33 FROST, S. W. (1959) *Insect life and insect natural history.* 2nd. Ed. N.Y.: Dover. 526 pp.

34 GILMOUR, DANCY (1961) *The biochemistry of insects.* London: Academic Press. 343 pp.

35 GOSTICK, K. C. & BAKER, P. M. (1966) Insecticide resistance in cabbage root fly. *Agriculture.* 73(2): 70–74.

36 GRESSITT, J. L. (1958) Zoogeography of insects. *Ann. Rev. Ent.* 3: 207–30.

37 GRESSITT, J. L. & WEBER, N. A. (1960) Bibliographic introduction to Antarctic and Sub-antarctic Entomology. *Pacific Insects.* **1**(4): 441–80.

38 ———— (1961) Supplement. *Pacific Insects.* 3(4):563–7.

39 GUNTHER, F. A. & JEPPSON, L. R. (1960) *Modern insecticides and world food production.* London: Chapman & Hall. 284 pp.

39a GUSTAFSON, J. F. (1950) The origin and evolution of the genitalia of the Insecta. *Microentomology.* **15**:35–67.

40 HARRIS, W. V. (1961) *Termites, their recognition and control.* London: Longmans Green. 187 pp.

41 HASKELL, P. T. (1961) *Insect Sounds.* London: Witherby (Aspects of Zoology series). 189 pp.

42 HINTON, H. E. (1946) A new classification of insect pupae. *Proc. zool. Soc. London.* **116**(2):282–328.

43 ———— (1958) The phylogeny of the Panorpoid Orders. *Ann. Rev. Ent.* **3**:181–206.

44 ———— (1962) Respiratory systems of insect egg-shells. *Science Progress.* **50**: 96–113.

45 ———— (1963) The origin and function of the pupal stage. *Proc. R. ent. Soc. Lond.* (A)**38**: 77–85.

46 HOCKING, B. (1960) Smell in insects: a bibliography with abstracts (to Dec. 1958) *Canadian E.P. Technical Rep.* **8**:266 pp.

47 HORSFALL, W. R. (1962) *Medical Entomology: Arthropods and human disease.* 467 pp.

48 HOTT, C. P. (1954) The evolution of the mouthparts of adult Diptera. *Stanford Univ. Bull.* **27**:(1951):137–8.

49 HUNTER-JONES, P. (1956) *Instructions for rearing and breeding locusts in the laboratory* [leaflet free from Anti-Locust Research Centre, London].

50 IMMS, A. D. (1957) *A general textbook of Entomology* 5th Edn. Revised by O. W. Richards and R. G. Davies. London: Methuen. 886 pp.

51 JANDER, R. (1963) Insect orientation. *Ann. Rev. Ent.* 8:95–114.

52 JOHNSON, C. G. (1962) Aphid migration. *New Scientist.* **15**:622–5.

53 ———— (1963) Physiological factors in insect migration by flight. *Nature.* **198**:423–7.

54 KENNEDY, J. S. (1939) The behaviour of the desert locust in an outbreak centre. *Trans. R. ent. Soc. Lond.* **89**:385–542.

55 ———— (1950) Host-finding and host-alternation in aphids. *Proc. 8th Int. Congr. Ent.* 423–6.

56 ———— (1951) The migration of the desert locust (*Schistocerca gregaria* Försk.) *Phil. Trans. R. Soc. Lond.* **235** (625):163–290.

57 KENNEDY, J. S. (1956) Reflex and instinct. *Discovery.* **17**(8):311–12.

58 KEY, K. H. L. (1950) A critique on the phase theory of locusts. *Quart. Rev. Biol.* **25**:363–407.

59 KIRKPATRICK, T. W. (1957) *Insect life in the tropics.* London: Longmans. 311 pp.

60 KNIPLING, E. P. (1955) Possibilities of insect control or eradication through the use of sterile males. *J. econ. Ent.* **48**:459–62.

61 KUHNELT, W. (1963) Soil-inhabiting Arthropods. *Ann. Rev. Ent.* 8:115–36.

62 KUIPER, J. W. (1962) The optics of the compound eye. *Soc. expl. Biol Symposia XVI: Biological receptor mechanisms.* 58–71.

63 LAPAGE, G. (1956) *Veterinary parasitology*. Edinburgh: Oliver & Boyd. 964 pp.

64 LEMCHE, H. (1941) *Origin of winged insects*. Copenh. Vid. Medd. 42 pp.

65 LHOSTE, J. (1962) Attractants for insect control. *S. P. A. N.* **5**:8–12.

66 LINDAUER, M. (1961) *Communication among social bees.* Oxford Univ. Press. 143 pp.

67 ———— (1967) Recent advances in bee communication and orientation. *Ann. Rev. Ent.* **12**:439–70.

68 LOAEZA, R. M. & CORONA, A. O. (1965) Esterlizaciòn de la Mosca doméstica con apholate. *Folia entomologica Mexicana.* **10**: 24 pp.

69 LOWER, H. F. (1959) The insect epicuticle and its terminology. *Ann. ent. Soc. Amer.* 52:381–5.

70 MACAN, T. T. (1962) Ecology of aquatic insects. *Ann. Rev. Ent.* 7: 261–88.

71 MANI, M. S. (1962) *Introduction to high altitude entomology*. London: Methuen. 302 pp.

72 ———— (1964) *The ecology of plant galls*. The Hague: Junk. 434 pp.

73 MANTON, S. M. (1953) Locomotory habits and the evolution of the larger arthropodean groups. *Evolution: Soc. exp. Biol. Symposium.* 7:339–76.

74 MARAMOROSCH, K. [Ed.] (1962) *Biological transmission of disease agents.*

75 ———— (1963) Arthropod transmission of plant viruses. *Ann. Rev. Ent.* 8:369–414.

76 MATHESON, C. (1962) British and world faunal estimates. *Ann. Mag. nat. Hist.* (**13**):705–11.

77 MATSUDA, R. (1958) On the origin of the external genitalia of insects. *Ann. ent. Soc. Amer.* 51:84–94.

78 ———— (1965) Morphology and evolution of the insect head. *Amer. Ent. Inst.* Ann Arbor, Michigan. **4**:334 pp.

79 NESBITT, H. H. (1941) A comparative morphological study of the nervous system of the Orthoptera and related orders. *Ann. ent. Soc. Amer.* **34**:51–81.

80 O'BRIEN, R. D. & WOLFE, L. S. (1964) *Radiation, radioactivity and insects.* N.Y. & London: Academic Press. 211 pp.

81 OLDROYD, H. (1958) *Collecting, preserving and studying insects.* London: Hutchinson. 327 pp.

82 ———— (1964) *The natural history of flies*. London: Weidenfeld & Nicolson. 324 pp.

83 PERCIVAL, MARY (1961) The flower and the bee. *Adv. Sci.* **18**:148–152.

84 POSPISIL, J. (1962) On visual orientation of the House fly (*Musca domestica*) to colours. *Acta Soc. ent. Czechoslovakia.* **59**:1–8.

85 PRINGLE, J. W. S. (1957) *Insect flight*. Cambridge Univ. Press. 133 pp.

86 QADRI, M. A. H. (1938) The life history of the cockroach, *Blatta orientalis* L. *Bull. ent. Res.* **29**:263–76.

87 RIBBANDS, C. R. (1953) *The behaviour and social life of honeybees.* London: Bee Research Assoc. 352 pp.

88 RIVNAY, E. (1962) Field crop pests in the Near East. *Monographiae Biologicae.* **10**:450 pp.

89 ———— (1964) The influence of man on insect ecology in arid zones. *Ann. Rev. Ent.* **9**:41–62.

90 ROSS, H. H. (1955) The evolution of the insect orders. *Ent. News.* **66**:197–208.

91 ———— (1966) A textbook of entomology. 3rd Edn. 519 pp.

92 ———— (1967) Evolution and past dispersal of the Trichoptera. *Ann. Rev. Ent.* **12**:169–206.

93 SCHNEIDER, D. (1964) Insect antennae. *Ann. Rev. Ent.* **9**:103–23.

94 SCOTT, J. P. (1958) *Animal behaviour.* Univ. Chicago Press. 281 pp.

95 SCUDDER, G. G. E. (1961) The comparative morphology of the insect ovipositor. *Trans. R. ent. Soc. London.* **113**:25–40.

96 SIMPSON, G. G. (1961) Principles of animal taxonomy. *Columbia Biol. Series.* **20**:247 pp.

97 SKAIFE, S. H. (1953) *African insect life.* London: Longmans. 387 pp.

98 ———— (1962) *The study of ants.* London: Longmans. 258 pp.

99 SOTAVALTA, O. (1947) The flight-tone (wing-stroke frequency) of insects. *Acta. ent. fenn.* **4**:1–117.

99a SOUTH, A. (1961) The taxonomy of *Entomobrya* (Collembola). *Trans. R. ent. Soc. Lond.* **113**:387–416.

100 SOUTHWOOD, T. R. E. (1962) Migration of terrestrial Arthropods in relation to habitat. *Biol. Revs.* **37**:171–214.

101 ———— (1966) *Ecological methods.* London: Methuen. 391 pp.

102 STRICKLAND, A. H. (1966) Some costs of insect damage and crop protection. *Pest Articles and News Summaries* (PANS). **12**:–57–72.

103 SUDD, J. H. (1967) *An introduction to the behaviour of ants.* London: Edward Arnold. 200 pp.

104 SWAN, L. A. (1964) *Beneficial insects.* N. York: Harper & Row. 429 pp.

105 SYMES, C. B., MUIRHEAD THOMPSON, R. C. & BUSVINE, J. R. (1962) *Insect control in public health.* Amsterdam and New York. 227 pp.

106 TAYLOR, L. R. (1960) The distribution of insects at low levels in the air. *J. anim. Ecol.* **29**:45–63.

107 THOMAS, JOAN G. (1963) *Dissection of the Locust.* London: Witherby. 72 pp.

108 THORPE, W. H. (1950) Plastron respiration in aquatic insects. *Biol. Revs.* **25**:344–90.

109 TREHERNE, J. E. & BEAMENT, J. W. L. [Eds.] (1965) *The physiology of*

the insect central nervous system. London & N. York: Academic Press. 227 pp.

110 TUXEN, S. L. (1964) *The Protura: a revision of the species of the world, with keys for determination*. Paris. 360 pp. [see also CLARKE, J. F. G.]

111 USINGER, R. L. [Ed.] (1956) *Aquatic insects of California, with keys to North American genera and Californian species*. Berkeley, California. 508 pp.

112 ——— & POVOLNY, D. (1966) The discovery of a possibly aboriginal population of the bed bug (*Cimex lectularius* L.). *Acta musei Moraviae: Scientiae naturales*. **51**:237–42.

113 UVAROV, B. P. (1961) Development of arid lands and its ecological effects on their insect fauna. *Arid Zones Res*. **18**:235–48.

114 von FRISCH, K. (1966) *The dancing bees*. 2nd Edn. London: Methuen. 198 pp.

115 WATERHOUSE, D. F. (1957) Digestion in insects. *Ann. Rev. Ent.* **2**:1–18.

116 WAY, M. J. (1963) Mutualism between ants and honeydew-producing Homoptera. *Ann. Rev. Ent.* **8**:307–44.

117 WELLINGTON, W. G. (1957) The synoptic approach to studies of insects and climate. *Ann. Rev. Ent.* **2**:143–62.

118 WEESNER, FRANCES M. (1960) Evolution and biology of the termites. *Ann. Rev. Ent.* 5:153–70.

118a WENNER, A. M. (1962). Sound production during the waggle-dance of the honey-bee. *Animal Behaviour* **10**:79–95.

119 WIGGLESWORTH, V. B. (1953) *The principles of insect physiology*. 5th Edn. London: Methuen. 554 pp.

120 ——— (1959) Metamorphosis, polymorphism, differentiation. *Sci. Amer.* **200**:100–10.

121 ——— (1959) The breathing-tubes of insects. *New Scientist*. 5:567–9.

122 ——— (1960) Fuel and power in flying insects. *New Scientist*. 8:101–4.

123 ——— (1962) Insect forms and evolution. *Discovery*. 22–7.

124 ——— (1964) *The life of insects*. London: Weidenfeld & Nicolson. 360 pp.

125 WILLIAMS, C. B. (1958) *Insect migration*. London: Collins (New Naturalist series). 235 pp.

126 WILSON, D. M. (1966) Insect walking. *Ann. Rev. Ent.* **11**:103–22.

127 WILSON, E. O. (1963) The social biology of ants. *Ann. Rev. Ent.* 8:345–68.

128 WRIGHT, R. H. (1958) The olfactory guidance of flying insects. *Canad. Ent.* **90**:81–9.

129 WYNNE-EDWARDS, V. C. (1962) *Animal dispersion in relation to social behaviour*. Edinburgh: Oliver & Boyd. 653 pp.

Index